AF902924

Population and Sustainable Development in India

Aalok Ranjan Chaurasia

Population and Sustainable Development in India

Springer

Aalok Ranjan Chaurasia
MLC Foundation and 'Shyam' Institute
Bhopal, Madhya Pradesh, India

ISBN 978-981-32-9211-6 ISBN 978-981-32-9212-3 (eBook)
https://doi.org/10.1007/978-981-32-9212-3

© Springer Nature Singapore Pte Ltd. 2020
This work is subject to copyright. All rights are reserved by the Publisher, whether the whole or part of the material is concerned, specifically the rights of translation, reprinting, reuse of illustrations, recitation, broadcasting, reproduction on microfilms or in any other physical way, and transmission or information storage and retrieval, electronic adaptation, computer software, or by similar or dissimilar methodology now known or hereafter developed.
The use of general descriptive names, registered names, trademarks, service marks, etc. in this publication does not imply, even in the absence of a specific statement, that such names are exempt from the relevant protective laws and regulations and therefore free for general use.
The publisher, the authors and the editors are safe to assume that the advice and information in this book are believed to be true and accurate at the date of publication. Neither the publisher nor the authors or the editors give a warranty, expressed or implied, with respect to the material contained herein or for any errors or omissions that may have been made. The publisher remains neutral with regard to jurisdictional claims in published maps and institutional affiliations.

This Springer imprint is published by the registered company Springer Nature Singapore Pte Ltd.
The registered company address is: 152 Beach Road, #21-01/04 Gateway East, Singapore 189721, Singapore

Contents

About the Author

Aalok Ranjan Chaurasia is M.Sc. from Kanpur University, Kanpur, India, Ph.D. from Jiwaji University, Gwalior, India, and holds Postgraduate Diploma in Population and Development from Jawaharlal Nehru University, New Delhi, India. He has participated in the United Nations Global Training Programme in Population and Development, 1990–1991 and has published more than 30 books and 120 research papers. A member of the International Union for the Scientific Study of Population and a life member of the Indian Association for the Study of Population, he also serves on the editorial board of the journal Demography India. His academic engagements include Director, State Population Resource Centre, Madhya Pradesh, India; Professor, Population Research Centre, Institute of Economic Growth, Delhi, India; and Project Director, Mother and Child Tracking System, Ministry of Health and Family Welfare, Government of India. He has also served as an expert member of the National Population Commission, Government of India.

Chapter 1
Introduction

Abstract The United Nations 2030 Sustainable Development Agenda is a plan of action for people, planet and prosperity with the central objective of promoting well-being of the current and the future generations. This chapter discusses the link between population and sustainable development with specific reference to India, the second most populous country of the world. The chapter also presents synopses of the chapters included in the monograph that focus on different dimensions of population, economy and environment in the Indian context.

The United Nations General Assembly adopted, in 2015, the post-2015 development agenda 'Transforming Our World: The 2030 Agenda for Sustainable Development' as the plan of action for people, planet and prosperity (United Nations 2015). The central objective of the agenda is to promote well-being of the current and the future generations. The agenda is summarised in 17 sustainable development goals (SDGs) and 169 targets to be achieved by the year 2030. The United Nations expects that the agenda will stimulate action at local, national and international levels in areas that are of critical importance for the humanity and the planet.

The term sustainable development has been defined in different ways in different contexts. The most popular conceptual foundation of sustainable development that is accepted universally is the concept of inter-generational equity. Following this concept, the United Nations World Commission on Environment and Development has defined sustainable development as the development that meets the needs of the present generation without compromising the ability of future generations to meet their own needs (United Nations 1987). The concept of inter-generational equity is closely associated with the concept of security of the present and future generations. Security has three dimensions—economic and social, environmental, and physical and personal. The economic and social security is related to meeting the needs of the present and future generations and is closely linked with the environmental security as the environment provides the natural resources necessary for sustaining life on the planet Earth and, at the same time, absorbs the wastes generated through use of natural resources and recycle them in the usable form. Finally, the physical and personal security is related to striking the balance between the economic and social security and the environmental security to ensure that efforts directed towards economic and

© Springer Nature Singapore Pte Ltd. 2020

A. R. Chaurasia, *Population and Sustainable Development in India*,
https://doi.org/10.1007/978-981-32-9212-3_1

social security are transformed into sustainable long-term practices. Physical and personal security is essentially a matter of governance—the manner in which the power is exercised in the management of natural resources to meet the needs of the people.

From the operational perspective, the concept of inter-generational equity or the issues related to economic, social, environmental, physical and personal security have been translated into the framework of socially inclusive and environmentally sustainable economic growth (Sachs 2015). This operational perspective may be viewed as the organising principle for meeting human development goals or the needs of the people while, at the same time, sustaining the ability of the environment to provide natural resources necessary for economic and social security through a socially inclusive process of change. The United Nations 2030 Sustainable Development Agenda follows this operational perspective to stimulate action in areas of critical importance for the humanity and the planet.

The challenge of achieving SDGs as articulated in the 2030 Sustainable Development Agenda is, however, a daunting one simply because the issues and concerns related to socially inclusive and environmentally sustainable economic growth are closely related to population transition or the change in the population stock—the size and the age composition. It is obvious that the task of reducing poverty and promoting human well-being and, at the same time, ensuring sustainability of the environment is inseparable from the change in the population stock. It is argued that efforts to promote sustainable development without addressing the population dynamics, therefore, are bound to fail (UNFPA 2012). In the context of sustainable development, population may be perceived as ‘the elephant in the room’. Although, issues and concerns related to population transition or the change in the population stock are out of the purview of the 2030 Sustainable Development Agenda, yet population transition is a decisive factor in the global environmental change and future human well-being. The change in the population stock affects all efforts directed towards meeting the needs of the people, including universal access to health, education for all, affordable housing, sanitation, water, food, energy, and so on. At the same time, the change in the population stock puts increasing pressure on the environment in terms of the increased demand of natural resources and in terms of the capacity of the environment to absorb wastes generated as a result of an increase in the natural resources use. The increased demand for natural resources to meet the needs of the increasing population and the resulting increase in the wastes generated contribute to climate change and challenge the capacity of the environment in sustaining life on the planet. According to the latest population projections prepared by the United Nations, it is most likely that the world population will increase from 7.383 billion in 2015 to 8.551 billion in 2030 and to 11.184 billion by the turn of the present century (United Nations 2017). This means that around 1.168 million people will be added to the world population during the 15 years between 2015 and 2030, while the net addition to the world population during 2015–2100 will be around 3.8 billion. This massive increase in the world population in the coming years will have serious implications for sustainable development processes.

At the same time, sustainable development may influence population transition significantly. It has been shown in a recent study that achieving SDGs by the year 2030 would lead to a global population of 8.2–8.7 billion by the turn of the current century (Abel et al. 2016). This means that achieving sustainable development goals is expected to result in curtailing the future growth of the world population by 2.0–2.5 billion which, in turn, will contribute significantly to improving personal, physical, social, economic and environmental security. The health- and education-related targets of the 2030 Sustainable Development Agenda will have direct and indirect consequences on future mortality and fertility trends which will have implications to the future population growth.

The relationship between population and sustainable development is essentially dynamic in nature on the scales of time and place. This means that the population and sustainable development relationship is always context-specific. A given set of concerns related to population and sustainable development, when addressed, may give way to a new set of sustainable development challenges and new concerns related to population transition that need to be incorporated in the sustainable development framework. On the other hand, characterisation of population and sustainable development relationship in one social, cultural, economic setting may be grossly out of context in other social, cultural and economic settings because of the sensitivity of both population and sustainable development to the prevailing social, cultural and economic context. This implies that the population and sustainable development inter-relationship must be studied through an empirical perspective and not through a normative, theoretical approach. The population and sustainable development inter-relationship can best be understood through a context-specific characterisation approach which is illustrative by design and is built upon the empirical evidence that is available.

Population transition and sustainable development in India matter for the global population transition and sustainable development simply because of the sheer size of India's population stock. India is estimated to have accounted for close to 18% of the world population in 2017. This proportion is very likely to remain unchanged during the period 2015–2030. At the same time, India, alone, will be responsible for more than 17% of the increase in the world population between 2015 and 2030 according to the medium but the most likely variant of the latest population projections prepared by the United Nations Population Division. These projections also suggest that after 2030, India's share to the world population is most likely to decrease so that by the end of the current century, India will be accounting for close to 14% of the world population.

On the other hand, the carbon dioxide emission in India is estimated to have increased by more than four times from 517 million tonnes in 1990 to 2166 million tonnes in 2015. During the same period, total primary energy use in the country increased from 307 million tonnes of oil equivalent to 882 million tonnes of oil equivalent in 2015. India accounted for about 3.6% of the global energy use and 2.6% of global carbon dioxide emissions in 1990. These proportions have since been estimated to have increased to 6.3 and 6.4%, respectively, in 2015. There is every possibility that there will be substantial increase in these proportions in the coming years because of continued population growth and expansion of the social and economic production system to meet the needs of the growing population.

In the recent past, India has recorded very rapid economic growth which has, however, been associated with increasing inequality. At the same time, there is little evidence to suggest that the rapid economic growth in India has translated into human progress that has implications for sustainable development. According to the World Bank, India is the home of the largest number of extremely poor people in the world—people living on <$1.90 per day (World Bank 2018). On the other hand, India had more deaths among children under five than any other country in 2015 according to a recent study (Chao et al. 2018). Similarly, it is estimated that around 15% of the total maternal deaths in the world happened in India in 2015 (WHO 2015). In terms of the human development index, India stands a poor 130 among 189 countries in 2018 (UNDP 2018), whereas India is the abode of the largest number of people living in multi-dimensional poverty (Oxford Poverty and Human Development Initiative 2018). Although India has made remarkable progress in terms of poverty reduction, yet the challenge of reducing the number of poor people in the country remains magnanimous because of the sheer size of India's population stock.

India's record in population transition has also not been very impressive despite the fact that India was the first country in the world to adopt an official policy on population and to launch an official programme of birth limitation way back in 1952. Population projections prepared by the United Nations suggest that it is most likely that India's population will increase by around 370 million from an estimated 1309 million in 2015 to 1679 million by the year 2061. It is only after 2061 that the population of the country is projected to decrease to 1517 million by the turn of the century (United Nations 2017). India's National Population Policy, announced in the year 2000, aimed at achieving the replacement fertility by the year 2010 and a stable population by the year 2045 at a level consistent with the requirements of sustainable economic growth, social development and environment protection (Government of India 2000). The goals of the National Population Policy 2000 could, however, not be achieved. The country is yet to achieve the replacement fertility, and the projections prepared by the United Nations suggest that the replacement fertility in India will most likely be achieved during 2025–2030 only—almost 25 years later than the target set in the National Population Policy 2000. It has been estimated that if the goals of the National Population Policy 2000 would have been met, the population of the country would have increased to around 1107 million by the year 2010 (Government of India 2000) against the estimated population of 1231 million in 2010 according to United Nations (2017). This means that slower than expected or planned population transition in the country had resulted in an addition of 124 million more people to the population of the country which has implications to sustainable development—economic growth, social inclusion and environmental protection. Between 2015 and 2030, more than 200 million people are expected to be added to the population of the country. This increase in the size of the population stock of the country will require substantial expansion of the social and economic production system so as to meet the needs of the growing number of people and eradicating poverty. The National Health Policy, announced in 2017, now aims at achieving replacement fertility at national and sub-national level by the year 2025 (Government of India 2017). It is obvious that the past, present and future population transition in the country will have implications for sustainable development process in general and realisation of the targets associated with SDGs that need to be taken into consideration while planning and programming for the 2030 Sustainable Development Agenda of the United Nations.

The above considerations constitute the rationale for the present monograph which is the third in the series of monographs devoted to population and development inter-relationships in India. The first monograph was published in 1999 (Ranjan 1999). It focused, primarily, on issues covered under the Global Training Programme in Population and Development sponsored by the United Nations Population Fund. The topics covered in the monograph included capital formation, savings and population growth; population growth and technology evolution; demand and supply of food; commercial energy use and impact of population and consumption on the environment; population policy and population transition; fertility transition; female reproductive age mortality; status of women; and urbanisation tends and patterns.

The second monograph was published in 2009 as the state of the population in India for India's National Commission on Population (Chaurasia and Gulati 2009). The monograph covered topics which had relevance to the National Population Policy 2000, including population growth and distribution, population and development, population policy, family planning, reproductive and child health, status of women, urbanisation and migration, population and development integration, and strategic options for the realisation of National Population Policy 2000 goals and objectives. The monograph also devoted a specific chapter on the progress of each state of the country towards the National Population Policy goals and objectives.

The present, third monograph, is directed towards analysing linkages between population dynamics and sustainable development in India in the context of the United Nations 2030 Agenda for Sustainable Development. It is a collection of papers each of which is related to a specific population and sustainable development concern that have implications to both population transition and sustainable development in the country. The papers included in the monograph illustrate the nature of population and sustainable development relationship under different social, cultural and economic contexts through an empirical perspective. The papers also serve as the framework for analysing population and sustainable development inter-relationship in different social, cultural and economic contexts or using a different set of empirical evidence.

The monograph comprises 10 chapters on different dimensions of population and sustainable development in India, in addition to this introduction. The first four chapters of the monograph focus on population transition processes in the country. The next two chapters are devoted to economic growth. The next three chapters focus on the social inclusion dimension of sustainable development, while the last chapter of the monograph analyses how population growth impacts upon sustainable development in the country. All chapters are based on the empirical evidence that has been derived from population and development-related data available from different sources, including population census, household surveys and official statistics of the government. The chapters included in the monograph present a composite population and sustainable development scenario in the country.

The second chapter of the monograph analyses population transition or the change in the population stock in the country during 1950–2015. The chapter follows a decomposition approach to analyse how different demographic drivers had contributed to the growth of the population of the country since independence in 1947. This approach takes into account both demographic transition and age structure transition in analysing the change in the population stock. The analysis suggests that the decrease in mortality in India has never been substantive enough to act as a catalyst to hastening the decrease in fertility. The analysis also indicates that population transition in the country has not been sensitive to different population and development policies followed by the country in the quest towards fertility reduction and population stabilisation. Moreover, the momentum of growth built in the population stock of the country has increasingly become dominant in deciding the rate of population growth than the level of fertility and mortality. The chapter concludes that the country requires appropriate policy response to address the current state of population transition so that the change in the population stock in the years to come remains in tune with the sustainable development processes, goals and objectives.

The third chapter of the monograph follows the methodology adopted in the first chapter to analyse future growth of the population stock of the country on the basis of the latest population projections prepared by the United Nations Population Division. The analysis suggests that the projected increase in India's population will be driven by its very large population size as the transition in fertility and mortality and the change in population age structure will contribute to the decrease in the population of the country in the years to come. The analysis also suggests that the effects of the decrease in the intrinsic population growth rate as a result of a decrease in fertility and mortality and the decrease in the natural population growth rate as a result of the change in population age structure on the future population growth will be offset by the momentum of growth resulting from the increase in the size of the population stock of the country. An important finding of the analysis is that slower than expected transition in fertility in the country in the past will have important implications for the future growth of the population of the country.

The fourth chapter of the monograph analyses fertility transition in India during the period 1991–2011. The analysis is based on a fertility transition index that takes into account transition in the two dimensions of fertility transition—birth planning and birth limitation. The analysis reveals that fertility transition efforts in India have largely remained confined to limiting births rather than planning births despite all talks of the paradigm shift at the policy level. There is some evidence of transition in birth planning during the 1990s but this transition either stagnated or reversed during the 2000s. The current pattern of fertility transition in India contributes little towards minimising the effect of momentum for future population growth in the population stock of the country. The analysis also indicates that birth limitation efforts in many states/Union Territories of the country are hardly relevant in the context of population stabilisation as replacement fertility has already been achieved in these states/Union Territories. The chapter argues that population stabilisation efforts in India must focus more on birth planning rather than birth limitation primarily to address the population growth attributed to the momentum for growth built in the population stock of the country.

Progress towards child survival in the country during 1971 through 2015 is the focus of the fifth chapter of the monograph. The analysis is based on the data available through India's official sample registration system. The chapter concludes that the progress towards child survival in the country has not been satisfactory during the period under reference so that the country could not be able to achieve the Millennium Development Goal of reducing the under-five mortality rate by two-thirds between 1991 and 2015 (MDG-4). The analysis reveals that child survival progress in the country slowed down after 1991–1995 and this slowdown appears to be the reason behind the country missing the Millennium Development Goal of reduction in child mortality. The chapter observes that if the pace of improvement in child survival probability observed during 1971–1975 through 1991–1995 would have been maintained during the period 1991–1995 through 2011–2015 also, then the country would have been able to achieve two-thirds reduction in the under-five mortality rate from the level that prevailed in 1990. The chapter also observes that, although, there has been some acceleration in the improvement in the child survival probability after 2001–2005, yet, the speed of improvement is too slow to achieve the targets set in the National Health Policy 2018 and under the United Nations 2030 Agenda for Sustainable Development.

The sixth chapter of the monograph analyses economic growth in the country during 1951–1952 through 2011–2012 in the context of the Five-year Development Plans that the country had since independence. The analysis follows the segmented or piecewise regression approach to analyse growth in the output of the economy during different Five-year Development Plans. The segmented regression analysis helps in decomposing the growth in the output of the economy into the growth attributed to the increase in the size of the economy and growth attributed to the acceleration in growth in successive Five-year Development Plans. The analysis reveals that almost 90% of the growth in the Indian economy since 1951–1952 has been the result of the increase in the size of the economy, whereas the growth acceleration has been frequently inconsistent. The analysis also reveals that around 40% of the growth in the Indian economy during 1950–1951 through 2011–2012 was due to the increase in the population stock whereas the increase in the per capita output of the economy accounted for about 60% of the growth of the output of the economy.

The seventh chapter of the monograph analyses the impact of population transition on economic growth in the country during the period 2001 through 2011. The analysis reveals that the contribution of the change in the population age structure or the demographic dividend on the increase in the per capita output has at best been marginal in India and nearly all the increase in the per capita output may be attributed to the increase in the productivity of workers or people engaged in social and economic production processes. The analysis suggests that the transition in fertility during the decade had not been substantial enough to have a telling impact on the increase in the per capita output of the economy. The analysis also reveals that demographic dividend varied widely across the states/Union Territories of the country and there is only one state—Goa—where the demographic dividend turned negative during the period under reference contributing towards slowing down instead accelerating economic growth. The diversity in demographic dividend across states/Union Territories of the country is expected as different states/Union Territories in India are at different stages of demographic transition.

The next three chapters of the monograph focus on the social inclusion component of sustainable development as characterised by the United Nations. Chapter 8 analyses human deprivation across the villages of the country following the capability approach and using the data mining technique. The analysis reveals that the level of human deprivation at the village level is determined by the defining characteristics of the village population which are largely exogenous to the social and economic production system. The primary reason for high to very high deprivation faced by the people living in the villages of the country is found to be very limited opportunities for the village people to participate in the social and economic productive activities at the village level. At the same time, the analysis highlights distinct regional patterns in the deprivation faced by the village people. The analysis also reveals that the three domains of the human capability framework—endowment, productive capacity and participation opportunity—are not essentially interlinked at the village level.

The ninth chapter of the monograph focuses on child deprivation in India. The state-level analysis, presented in this chapter, suggests that child deprivation in India and in its constituent states remains highly pervasive. A substantial proportion of Indian children continue to be devoid of even the basic services and facilities that are necessary for their well-being—survival, growth, development and protection. The analysis also suggests that the proportion of children facing deprivation varies across states and, within states, by residence and social class. Moreover, the within-country differences in the proportion of children facing one or the other form of deprivation appears to be quite pervasive in terms of strength and persistence. The chapter argues that such a situation, obviously, cannot be accepted in the context of the best interests of children to which the country is committed, at least at the policy level. The analysis suggests that addressing the residence and social class inequality that is so pervasive in India can go a long way in promoting and securing child well-being in the country.

The tenth chapter of the monograph analyses female work participation in the villages of the country on the basis of the data available through the 2011 population census. The analysis indicates that the level of female participation in the village economy in India remains far from satisfactory. There are very few villages in the country where female participation in the village level productive activities may be termed as satisfactory. More specifically, there is only a nominal participation of females in household-level productive activities in all but a few villages, and one probable reason is that there are very limited opportunities for household-level productive activities in the village economy. The analysis also suggests that the level of female participation in the village economy is influenced by the social class structure, gender composition of the village population, the level of female education and the level of fertility in the village in addition to the size of the village population. A revealing finding of the analysis is that the level of female participation in village economy is negatively associated with the level of female education in the village.

This negative association contradicts the widely held argument that female education increases opportunities for the participation of women in social and economic productive activities. It appears that opportunities of participation of educated females in the economy of Indian villages are very limited.

The last chapter of the monograph analyses the impact of population growth on environmental sustainability, which is one of the three dimensions of sustainable development as identified in the 2030 Sustainable Development Agenda of the United Nations. The analysis reveals that population growth in the country had a substantial impact on both increase in the primary energy use and increase in the carbon dioxide emission in the country in the past, and the likely impact of the future population growth on the environment will be even greater. As such, the analysis calls for considering population dynamics while customising the 2030 Sustainable Development Agenda to the Indian context. The 2030 Sustainable Development Agenda is essentially a global development perspective which, as recommended by the United Nations, needs to be customised at the country level after incorporating development priorities and concerns specific to the country.

The analysis presented in this monograph highlights the need of integrating population factors in the sustainable development planning and programming processes. This integration requires putting in place a lens and mirror mechanism to understand the complex, yet very diverse inter-relationship between population and development based on the sustainability perspective. The mechanism emphasises that all sustainable development programmes and efforts must be viewed through the population lens to understand how these programmes and efforts have influenced the population stock—the size and the age composition. People are at the centre of all sustainable development efforts. The change in the population stock, therefore, must be at the core of all challenges and opportunities for sustainable development. It is imperative that sustainable development programmes and efforts must account for and address population factors.

On the other hand, integration of population factors in sustainable development planning and programming must also ensure that all efforts directed towards population transition—reduction in fertility and mortality and population stabilisation—must have reflections in terms of the improvement in the quality of life of the people by expanding individual opportunities and choices, empowering women, combating poverty and developing human capabilities, and functioning and strengthening social security systems. This requires pro-active planning and programming for population transition rather than reactive response to the prevailing population situation. Given the very diverse social, economic, cultural and environmental milieu of the country India, it is imperative that such a mechanism must be institutionalised at the grassroots level—the interface with the people. This is necessary to take into account the local context of both population transition and sustainable development.

References

Abel GJ, Barakat B, Samir KC, Lutz W (2016) Meeting the sustainable development goals leads to lower world population growth. Proc Natl Acad Sci 113(50):14294–14299

Chao F, You D, Pederson J, Hug L, Alkema L (2018) National and regional under-five mortality rate by economic status for low-income and middle-income countries: a systematic assessment. Lancet Global Health 6:e535–547

Chaurasia AR, Gulati SC (2009) India: The State of Population 2007. National Commission on Population and Oxford University Press, New Delhi

Government of India (2000) National population policy 2000. Ministry of Health and Family Welfare, New Delhi

Government of India (2017) National health policy 2017. Ministry of Health and Family Welfare, New Delhi

Oxford Poverty and Human Development Initiative (2018) Global Multi-dimensional poverty index 2018. The most detailed picture to date of the world's poorest people. Oxford University, Oxford

Ranjan A (1999) Population and development: the indian perspective. Datia, Madhya Pradesh, India, 'Shyam' Institute. ISBN 1-58112-844-4

Sachs JD (2015) The age of sustainable development. Colombia University Press, New York

United Nations Development Programme (2018) Human development indicators and indexes: 2018 statistical update. United Nations Development Programme, New York

United Nations (1987) Report of the world commission on environment and development. United Nations, New York

United Nations (2015) Transforming our world: the 2030 Agenda for sustainable development. United Nations, New York

United Nations (2017) World population prospects. The 2017 revision. Department of Economic and Social Affairs. Population Division, New York

United Nations Population Fund (2012) Population matters for sustainable development. United Nations Population Fund, New York

World Health Organization (2015) Trends in maternal mortality rate: 1990-2015. Estimates by WHO, UNICEF, UNFPA, World Bank Group and United Nations Population Division. Geneva, World Health Organization

World Bank (2018) Piecing together the poverty puzzle. The World Bank, Washington DC

Chapter 2
Population Transition 1950–2015

Abstract This chapter analyses population transition in India during 1950–2015. It is estimated that the increase in India's population had accounted for almost 20% of the increase in the world population during the period 1950 through 2015. The chapter analyses how transition in fertility and mortality and the associated change in population age structure has contributed to the change in the population stock—size and age structure. It is argued that India should appropriate its policy response to the current state of population transition so that the change in the population stock in the coming years remains in tune with sustainable development processes.

2.1 Introduction

India is the second most populous country in the world, next only to China. In 2015, India's population accounted for almost 18% of the world population (United Nations 2017). The increase in India's population is estimated to have accounted for almost 20% of the increase in the world population during 1950–2015. As per the 1951 population census, country's population was enumerated to be around 361 million which increased to around 1210 million according to the 2011 population census. On the other hand, estimates prepared by the United Nations Population Division suggest that India's population increased from around 376 million in 1950 to more than 1309 million in 2015. Projections prepared by the United Nations Population Division suggest that India is most likely to surpass the population of China to become the most populous country in the world by the year 2025 and the population of the country is most likely to continue to increase up to the year 2062 when it will cross the 1678 million mark despite continued decrease in fertility and mortality (United Nations 2017). Most of the increase in the population in the next 50 years will be due to the in-built momentum for growth in the population which is the result of the past fertility and mortality dynamics (Chaurasia 2017). India continues to be in the middle of population transition with an average annual population growth rate of around 1.3% per year; total fertility rate above the replacement level; and expectation of life at birth at around 68 years. During 2010–2015, around 80 million people were added to the population of the country.

© Springer Nature Singapore Pte Ltd. 2020

A. R. Chaurasia, *Population and Sustainable Development in India*,
https://doi.org/10.1007/978-981-32-9212-3_2

It is well known that population transition may have important implications to economic growth resulting from the decrease in fertility and associated decrease in the dependency ratio leading to an increase in the share of the population concentrated in the working ages or the ratio of the working age to the non-working age population (Bloom and Williamson 1998; Bloom and Canning 2004, 2005; Bloom et al. 2000, 2002; Cai and Wang 2005; Mason 2007). The increased concentration of the population in the working age suggests that even if there is no change in the output per worker, labour force participation rates and unemployment rates, an increase in the share of the working age population will lead to an increase in output per capita which is attributed purely to demographic changes and hence the name demographic dividend. The impact of population transition on economic growth is, however, policy-dependent (Mason 2007). A growth strategy that can produce additional jobs to productively utilise rapidly growing workforce is necessary to encash demographic changes in economic terms.

The available evidence suggests that the demographic dividend in India has never been large enough to have a telling impact on the economic growth (Chaurasia and Gulati 2008). It is however argued that much of India's potential demographic dividend lies ahead (Golley and Tyers 2012). Choudhry and Elhorst (2010) have estimated that population dynamics explained only 39% of the growth in per capita GDP in India. It is expected that future population dynamics will have a positive effect on the economic growth in India (Choudhry and Elhorst 2010). It has also been emphasised that the demographic dividend will offer the single biggest hope for India to catch up with China in the coming years (Kelkar 2004).

This chapter analyses population transition in India during the 65 years between 1950 and 2015 to identify distinctive features of the population transition path followed by the country. More specifically, the chapter analyses how transition in fertility and mortality and the associated change in population age structure have contributed to the change in the population stock—size and age structure. There are many studies that have analysed population dynamics in India in different contexts, including economic growth (Bloom et al. 2006; Da Vanzo et al. 2011; Bloom 2011; Golley and Tyers 2012), migration (White and Subedi 2008), savings and economic growth (Joe et al. 2015), macroeconomics (Wolf et al. 2011), political economy (Desai 2003), competitiveness (Schmidt 2011), and so on. However, there are a few studies that have analysed population dynamics in the context of population transition—transition in fertility, mortality and age composition (Coale 1983; Roy 1993; Adlakha and Banister 1995).

The chapter is organised as follows: The next section outlines the methodology used to analyse population transition. Section three describes the data source. The analysis is based on the latest population estimates prepared by the United Nations Population Division. Section four presents an outline of population growth in India during 1950 through 2015. Section five decomposes the change in the natural population growth rate in the country into the change attributed to transition in fertility, mortality and population age structure. Section six decomposes the increase in the population into the increase attributed to fertility, mortality, age composition and population size. Transition in the age composition of the population has been analysed in section seven of the chapter. The findings of the analysis are discussed in the last section of the chapter through the policy perspective.

Analysis of the key drivers of population growth has always been an area of interest to demographers. A number of studies have analysed future population growth under a set of assumptions about future trends in factors that affect population growth (Andreev et al. 2013; Bongaarts 1994, 2009; Bongaarts and Bulatao 1999; Frejka 1973, 1981). For example, Andreev (2013) has compared four variants of population projections to estimate how future changes in different components of the population growth will influence the future growth of the world population. Such analyses can inform policies and programmes aimed at balancing impending demographic changes and social, economic, health and environmental objectives (Bongaarts 1994; Bongaarts and Bulatao 1999). Notestein (1960) has used this approach to explore how changes in different factors affecting population growth were related to the change in the population stock of the United States of America during 1930–1955.

There are also studies that have analysed demographic factors responsible for the change in the population stock in the past (Smith and Ahmed1990; Horiuchi 1992, 1993; Kandiah and Horiuchi 1995). Horiuchi (1992, 1993) has analysed the stagnation in the decline of the world population growth rate during the 1980s, whereas Kandiah and Horiuchi (1995) have analysed trends and prospects of the world population growth. They identified three main factors behind stagnation: age structure, fertility trends in China and India and the fact that in many developing countries, the decrease in fertility slowed down during the 1980s compared to 1960s and 1970s. Alternatively, population growth can also be considered as the sum of the increase in population of different ages (Horiuchi 1995). This approach reflects the cohort replacement view of population growth against the entrance–exit view of population growth according to which population is a group of individuals with two entrances (birth and immigration) and two exits (death and emigration). Horiuchi (1995) has used this approach to analyse the determinants of the change in population growth rates in Sweden. He has shown that the growth rate of the population is a weighted sum of past changes attributed to the change in population size, age distribution, fertility, mortality and migration and has named this decomposition as the retrospective decomposition of growth rate distinguishing it from the synchronous decomposition based on the growth balance equation of relative population change. The difference between the two is rooted in the period versus cohort approach of demographic analysis.

A consistent theme in research on population transition is related to the change in the population stock as population moves from the pre-transition phase to the post-transition phase. A synthetic index that reflects the change in population size during transition from pre-to post-transition phase is the transition multiplier, which is the number by which population is multiplied when it moves from the pre-transition phase to the post-transition phase (Chesnais 1979, 1986). The population multiplier can be calculated theoretically as well as empirically. The theoretical approach involves the approximation either through an analytical model (Keyfitz 1971) or by numerical simulation (Frejka 1973). Empirical approach, on the other hand, involves examination of the historical evidence complemented by long range population projections (Chesnais 1990). On the other hand, the change in population age composition during the movement from pre- to post-transition phase is measured in terms of the change in the dependency ratio or, equivalently, the change in the potential support ratio. The dependency ratio is defined as the ratio of the number of children (0–14 years) and the number of old persons (60 or 65 years and above) to the number of working age population (15–60 or 15–64 years), while the potential support ratio is reciprocal to the dependency ratio. Population transition has three phases with specific characteristics of population age composition. At the early stages of transition, the dependency ratio is high or the potential support ratio is low because of a high proportion of the young population—population below 15 years of age. At the middle stages of transition, the dependency ratio decreases or the potential support ratio increases and at the later stage of transition, the dependency ratio is again high or the potential support ratio is low because of a high proportion of the old population—population aged 60 or 65 years and above (Williamson 1977). During the period when the dependency ratio decreases or the potential support ratio increases, there is demographic window of opportunity which may be reaped to accelerate economic growth.

2.2 Methodology

Population transition refers to the changes in the population stock or in the size and the age composition of the population as a result of change in factors that affect the stock (Schoen 2002). A live birth results in an increase in the population while a death results in a decrease. The birth always increases the population at age zero while the death can occur at any age and, therefore, change in the size of the population is also associated with the change in its age composition. The population may also decrease or increase and its age composition may change if a member of the population leaves it or if a person from outside enters the population. If P(0) denotes the population at time 0; B(0, t) denotes the number of live births during the time interval (0, t); D(0, t) denotes the number of deaths; I(0, t) denotes the number of in-migrants; and O(0, t) denotes the number of out-migrants, then the population at time t, P(t) can be obtained from the simple identity known as the balancing equation of population change (Preston et al. 2001:2).

$$P(t) = P(0) + B(0,t) - D(0,t) + I(0,t) - O(0,t) \quad (2.1)$$

If inward and outward movement of individuals is ignored, then the change in the size of the population during a given period depends upon the number of births and the number of deaths during that period so that

$$P(t) - P(0) = B(0,t) - D(0,t) \quad (2.2)$$

Dividing both sides of Eq. (2.2) by the person-years lived during the interval (0, t), PY(0, t), we get

$$r = b - d. \tag{2.3}$$

where r is termed as the (crude) rate of natural increase (Preston et al. 2001) or the natural population growth rate, b is the (crude) birth rate and d is the (crude) death rate. Thus, there are four factors in the absence of migration that determine the change in the population stock: (1) initial population size; (2) birth rate; (3) death rate; and (4) the period during which given birth and death rates apply (Vallin 2006). When $r = 0$, the size of the population stock remains unchanged; when $r > 0$, the population stock increases; and when $r < 0$, the population stock decreases. The larger the magnitude of r, the larger is the change in the population stock. In this way, r serves as a useful indicator of population transition. The demographic transition theory (Notestein 1945) says that as population moves from the pre-transition phase characterised by high birth rate and high death rate to the post-transition phase characterised by low birth rate and low death rate, the size of the population increases, and its age composition changes. The reason is that there is a time gap between the onset of the decrease in the death rate and the onset of the decrease in the birth rate so that the natural population growth rate, r, first increases and then decreases.

The prime determinant of the birth rate is the actual birth performance or fertility of the individual woman. However, only women in the reproductive age group (15–49 years) are biologically capable of producing births. Moreover, the probability of conception also varies by the age of the woman. In other words, the birth rate has two components, one age independent and the other influenced by population age composition. If f denotes the birth rate independent of age, then we can write

$$b = f * (b/f) = f * ab \tag{2.4}$$

where $ab = b/f$ is a multiplier which reflects the effect of age composition on the birth rate. Similarly, the death rate also has two components. If l denotes the age-independent death rate, then, we can write

$$d = l * (d/l) = l * ad \tag{2.5}$$

Equation (2.4) may now be written as

$$r = f * ab - l * ad$$

Now, following Kitagawa (1955), r can be decomposed into two components as follows:

$$r = (f - l) * (ab + ad)/2 + ((f + l) * (ab - ad)/2 = i + a \tag{2.6}$$

The first component on the right of Eq. (2.6) is the intrinsic component which is determined by the prevailing regimes of fertility and mortality as measured by the age-independent birth rate, *f*, and the age-independent death rate, *l*, and may be termed as the intrinsic rate of population growth (Espenshade 1975). The second component is the age component, *a*, which reflects the age composition effects on the birth rate, *ab*, and the age composition effects on the death rate, *ad* and hence on *r*. It is obvious from Eq. (2.6) that *r* will not be zero, even if $f = l$ or when $i = 0$. Obviously, the change (increase or decrease) in the population stock depends not only on the change in fertility and mortality but also on the change in population age composition. The age component of the natural population growth rate is closely related to the concept of population momentum (Keyfitz 1971; Schoen and Kim 1991) which, in its narrowest formulation, refers to the fact that a population closed to migration may continue to increase even when fertility decreases to the replacement level (Preston et al. 2001). Population momentum measures the cumulative contribution of age composition to population change in all future years after fertility has been set at the replacement level. It is usually measured in relative terms as the ratio of the size of the long-run stationary population to that of the observed population when replacement fertility is first attained (Espenshade and Tannen 2015) and can be both greater than or less than the limiting value of 1. When the age composition of the population is younger than the age composition of the stationary population, the momentum factor exceeds 1. In this case, the population continues to increase even after the replacement fertility is imposed. If the age composition of the population is older than that of the stationary population, the momentum factor is less than 1. In this case, the actual population will decrease, instead increase, when the replacement fertility is imposed. When the population age composition is the same as that of the stationary population, the momentum factor is 1. In this case, the natural population growth rate is determined solely by the intrinsic population growth rate (Preston and Guillot 1997). It is, however, clear from Eq. (2.6) that the concept of momentum is not confined to imposing replacement fertility only. It applies to any change in the population stock as the change in the stock depends not only upon the change in fertility and mortality but also upon the change in the age composition which is actually the legacy of past fertility and mortality dynamics (Preston et al. 2001).

Equation (2.6) suggests that the change in the natural population growth rate, *r*, should be analysed in terms of the change in the age-independent birth rate, *f*, age-independent death rate, *l*, age composition effects on the birth rate, *ab*, and age composition effects on the death rate, *ad*. Now

$$\nabla r = r_2 - r_1 = (b_2 - d_2) - (b_1 - d_1) = (b_2 - b_1) + (d_2 - d_1) = \nabla_b - \nabla_d \quad (2.7)$$

and

$$\nabla_b = (b_2 - b_1) = f_2 * ab_2 - f_1 * ab_1$$
$$= [(f_2 - f_1) * (ab_1 + ab_2)/2] + [(ab_2 - ab_1) * (f_1 + f_2)/2]$$

$$= \partial_f + \partial_{ab} \tag{2.8}$$

Similarly

$$\begin{aligned}\nabla_d = (d_2 - d_1) &= l_2 * ad_2 - l_1 * ad_1 \\ &= [(l_2 - l_1) * (ad_1 + ad_2)/2] + [(ad_2 - ad_1) * (l_1 + l_2)/2] \\ &= \partial_l + \partial_{ad}\end{aligned} \tag{2.9}$$

Substituting (2.8) and (2.9) in (2.7), we get

$$\begin{aligned}\nabla r &= (\partial_f + \partial_{ab}) - (\partial_l + \partial_{ad}) \\ &= (\partial_f - \partial_l) + (\partial_{ab} - \partial_{ad}) = \nabla_i + \nabla_a\end{aligned} \tag{2.10}$$

The first term on the right of Eq. (2.10) is determined by the change in the age-independent birth rate, *f*, and the age-independent death rate, *l*, or by the change in the intrinsic rate of population growth, while the second term is determined by the change in the age composition of the population as it affects the birth rate and the death rate. It is also obvious from Eq. (2.10) that ∇r will not be zero but will be equal to ∇_a even when there is no change in fertility and mortality so that $\nabla_i = 0$. Equation (2.10) takes into account both change in the intrinsic rate of population growth and change in the population age composition in explaining the change in the population stock. It addresses the controversial issue of which demographic indicators are more efficient in analysing population transition. In the absence of migration, the logical choice of the indicator reflecting population transition is the natural population growth rate which is the difference between the birth rate and the death rate. However, both birth and death rates are crude measures of fertility and mortality which are affected, in addition to the level of fertility and mortality, by the population age and sex composition. Using only crude rates to measure population transition, therefore, may obscure the change in fertility and mortality. Birth and death rates may stay at a relatively high level despite low levels of fertility and mortality because population age composition contributes to keep them high. On the other hand, the fertility indicator like the total fertility rate and the mortality indicator like expectation of life at birth have limitations in analysing population transition. First, both the indicators are based on hypothetical rather than actual population. Second, change in the total fertility rate and in the expectation of life at birth does not necessary lead to change in the birth rate and the death rate, and hence in the natural population, growth rates as birth and death rates are also influenced by changes in population age and sex composition. Analysis of population transition requires that changes in both fertility and mortality and changes in population age composition should be taken into account. Equation (2.10) provides this analytical framework. It essentially weighs the differences in the levels of fertility and mortality by the average levels of the age structure effects on birth and death rates, while the difference in the age structure effects on birth and death rates are weighed by average levels of fertility and mortality. There are other decomposition procedures also but they introduce a residual or interaction term which is often difficult to interpret (Preston et al. 2001).

Decomposing the change in the natural population growth rate, *r,* into its components has policy relevance also. If the change in the age component accounts for, say, 80% of the change in the natural population growth rate then there is little relevance of policies and programmes directed towards reducing fertility and mortality in the context of population transition. In such a scenario, population policies and programmes should be directed towards addressing the age component of the natural population growth rate, which is the result of the past fertility and mortality dynamics (Bongaarts 1994). For example, Bourgeois-Pichat and Taleb (1970) have shown that the target set by Mexico to achieve zero population growth rate by 2000 was unrealistic because the age composition of the population of Mexico was too conducive to positive growth of the population.

Application of Eq. (2.10) requires measurement of the age-independent birth rate, *f*, and the age-independent death rate, *l*. A measure of *f* can be derived from the total fertility rate (TFR) which is actually the sum of the age-specific birth rates of reproductive age women. Dividing the total fertility rate (TFR) by 35—the length of the reproductive period—gives the age-independent birth rate per woman of reproductive age. Finally, multiplying the age-independent birth rate per woman of reproductive age by the proportion of reproductive age women (*w*) in the population gives an estimate of the age-independent birth rate, *f*. Thus, the age-independent birth rate, *f*, may be calculated from the total fertility rate as $f = \mathrm{w} \times (\mathrm{TFR}/35)$. It may be noticed that *f* is actually a scalar multiple of the total fertility rate (TFR). Horiuchi (1991) has shown that the ratio of the birth rate to TFR is a measure of the age structure effects on the birth rate. As such, the ratio b/f is also a measure of the age structure effects on the birth rate.

Similarly, the expectation of life at birth (e_0) is a measure of mortality which depends upon age-specific probabilities of death only and is not influenced by the age composition of the population. The reciprocal of e_0 which is also known as the life table death rate may therefore be taken as a measure of the age-independent death rate, *l* and the ratio d/l may be argued to be a measure of the age composition effects on the death rate in the same manner as the ratio b/f is a measure of the age composition effects on the birth rate. It may, however, be noticed that interpretation of the ratio d/l as a measure of the age composition effects on the death rate is not as straightforward as the interpretation of the ratio b/f as a measure of the age composition effects on the birth rate (Horiuchi 1992). It depends on the stage of mortality transition (United Nations 2006). At the early stage of transition, the decrease in mortality is primarily confined to infancy and early childhood which exerts an influence on population change which is similar to the increase in fertility and, therefore, initial stages of mortality transition tend to make the population as a whole, grow younger, even as the probability of survival to middle and older ages increase. As mortality transition advances, the decrease in mortality is increasingly confined to older ages so that the late stage of mortality transition tends to make the population as a whole, grow older, by increasing the proportion of the old in the population.

We have used two indicators to measure the change in population age structure resulting from population transition: (1) the potential support ratio or the ratio of the working age population (15–60 or 15–64 years) to the non-working age population (0–14 years and 60 or 65 years and above) which is the reciprocal of the conventional dependency ratio; and (2) the difference between the growth rate of the working age population and the growth rate of the non-working age population in a given time period. The two indicators have purposely been selected to capture the transition in the population age structure in place of the proportionate age distribution of the population. The reason is that the two indicators are also used to quantify the demographic dividend that is generated as a result of population transition and which has implications for economic growth. The demographic dividend is accrued in a given time period when the growth rate of the working age population is relatively more rapid than the growth rate of the total population during that time period so that the difference between the two growth rates is positive. When the growth rate of the working age population is slower than the growth rate of the non-working age population, the difference between the two growth rates is negative and no demographic dividend is accrued. It is also obvious that the larger is the difference between the growth rate of the working age population and the growth rate of the non-working age population, the larger is the demographic dividend generated as a result of population transition, and hence the larger is the potential contribution of population transition in terms of accelerating economic growth even if other determinants of economic growth remain unchanged.

The analysis is based on the latest estimates of population and selected demographic indicators for India prepared by the United Nations Population Division for the period 1950–2015, including estimates of net migration during different quinquennials of the period 1950–2015. Using these two data, we have estimated India's population without adjustment for net migration. A comparison of population estimates adjusted for net migration and not adjusted for net migration suggests around 6 million people migrated out of India during 1950–2015. The present analysis uses population estimates which have not been adjusted for net migration.

Estimates of India's population are also available from the Registrar General and Census Commissioner of India. These estimates are based on decennial population census. We have not used these estimates for two reasons. First, these estimates do not include those who have migrated out of the country. Second, there is under-enumeration in the population census in India. The post-enumeration check after the 2011 population census revealed an under-enumeration of 2.3% in the census count (Government of India 2014). However, the difference between UN estimates and estimates based on decennial population census is small. Moreover, census estimates are not available for different quinquennials of the period 1950–2015. Therefore, charting of the population transition path followed by the country during the period under reference requires interpolation of different data related to population growth and its determinants which has not been preferred in the present analysis. Estimates prepared by the United Nations Population Division provide a consistent data set for analysing population transition in the country.

2.3 Population Growth in India

United Nations Population Division has estimated that India's population increased from around 376 million in 1950 to more than 1309 million in 2015 (Table 2.1). If the net migration is also taken into account then India's population is estimated to have increased to almost 1315 million by 2015 which means that around 6 million people migrated out of the country during 1950–2015. The net addition to the population of the country increased during every quinquennial till 2000 so that during 1995–2000 alone, more than 93 million people were added to the population of the country. However, the net addition to the population in a quinquennial started decreasing after 2000 and, during 2010–2015, the net addition to the population was estimated to be less than 81 million (Fig. 2.1).

During the same period, the crude birth rate in the country decreased from around 43.6 births per 1000 population during 1950–1955 to around 20 births per 1000 population during 2010–2015, whereas the crude death rate decreased from 26.8 deaths per 1000 population during 1950–1955 to 7.3 deaths per 1000 population during 2010–2015 according to the estimates prepared by the United Nations Population Division. On the other hand, the total fertility rate decreased from around 5.90 births per woman of reproductive age (15–49 years) during 1950–1955 to around 2.44 births per woman of reproductive age during 2010–2015, whereas the expectation of

Table 2.1 Population growth in India, 1950–2015

Year	Population (million)		Birth rate (0/00)	Death rate (0/00)	Total fertility rate	Expectation of life at birth (years)
	Adjusted for migration	Not adjusted for migration				
1950	376.325	376.325				
1955	409.269	409.375	43.626	26.786	5.903	36.620
1960	449.481	449.74	42.646	23.831	5.896	39.656
1965	497.702	498.047	41.509	21.095	5.890	42.732
1970	553.579	554.269	39.862	18.461	5.723	46.024
1975	621.302	619.849	38.504	16.127	5.410	49.393
1980	696.784	694.173	36.770	14.109	4.974	52.546
1985	781.667	778.473	35.682	12.747	4.682	54.933
1990	870.133	866.895	33.154	11.627	4.266	56.732
1995	960.483	957.797	30.164	10.212	3.833	59.182
2000	1053.051	1051.082	27.699	9.105	3.483	61.566
2005	1144.119	1144.101	25.371	8.407	3.141	63.543
2010	1230.981	1233.877	22.922	7.810	2.796	65.566
2015	1309.054	1314.528	19.992	7.326	2.438	67.575

Source United Nations (2017)

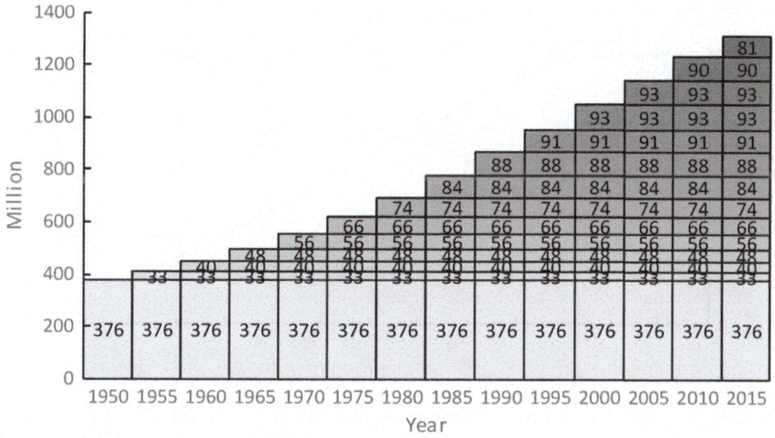

Fig. 2.1 Population growth in India, 1950–2015

life at birth increased from around 36.6 years during 1950–1955 to around 67.6 years during 2010–2015. The total fertility rate in the country still appears to remain above the replacement level fertility of 2.1 births per woman of reproductive age according to the estimates prepared by Unites Nations Population Division (Figs. 2.2 and 2.3).

The average annual natural population growth rate in the country was low during 1950–1955 but increased till 1980–1985 when the population increased at an average annual rate of more than 2.29% (Table 2.2). The average annual population growth rate started decreasing after 1980–1985 but the decrease gained momentum only after 2000–2005. On the other hand, the intrinsic population growth rate increased till 1965–1970 to peak at 1.54% per year whereas the age component of the natural population growth rate always contributed to increase in the natural population

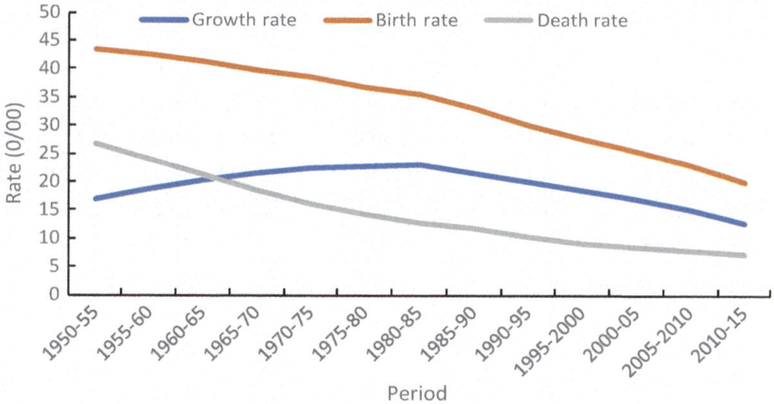

Fig. 2.2 Population growth rate, birth rate and deaths rate in India, 1950–2015

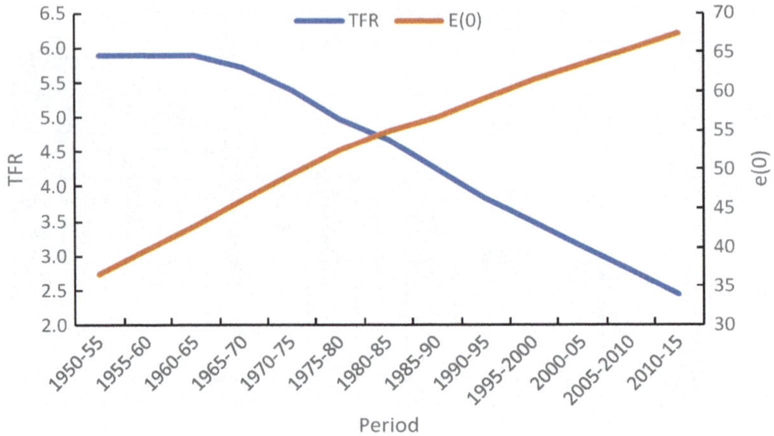

Fig. 2.3 Total fertility rate and expectation of life at birth in India, 1950–2015

Table 2.2 Components of natural population growth rate in India, 1950–2015

Period	Natural population growth rate	Components of natural population growth rate		Age independent		Age composition effects on	
		Intrinsic component	Age component	Birth rate	Death rate	Birth rate	Death rate
	(0/00)	(0/00)	(0/00)	(0/00)	(0/00)	*ab*	*ad*
1950–1955	16.840	13.454	3.386	40.762	27.307	1.070	0.981
1955–1960	18.815	14.469	4.346	39.686	25.217	1.075	0.945
1960–1965	20.414	15.113	5.301	38.514	23.402	1.078	0.901
1965–1970	21.401	15.388	6.013	37.116	21.728	1.074	0.850
1970–1975	22.377	15.241	7.136	35.487	20.246	1.085	0.797
1975–1980	22.661	14.064	8.597	33.095	19.031	1.111	0.741
1980–1985	22.935	13.227	9.708	31.431	18.204	1.135	0.700
1985–1990	21.527	11.218	10.309	28.844	17.627	1.149	0.660
1990–1995	19.952	9.396	10.556	26.293	16.897	1.147	0.604
1995–2000	18.595	8.196	10.399	24.439	16.243	1.133	0.561
2000–2005	16.965	6.735	10.230	22.472	15.737	1.129	0.534
2005–2010	15.112	4.962	10.150	20.214	15.252	1.134	0.512
2010–2015	12.665	2.973	9.693	17.771	14.798	1.125	0.495

Source Author's calculations

growth rate. The contribution of the age component increased till 1995 and decreased thereafter, although the decrease has been very slow. The relative contribution of the intrinsic growth and the age component to the natural population growth rate reversed over time. The age component accounted for almost 80% of the natural population growth rate during 2010–2015.

Table 2.2 also shows that age composition effects, in general, inflated the age-independent birth rate, f but deflated the age-independent death rate l. The age composition induced inflation in the age-independent birth rate f increased since 1950–1955 till 1985–1990 when it inflated the age-independent birth rate f by almost 15% but decreased thereafter, albeit slowly. During 2010–2015, the age composition of the population inflated the age-independent birth rate f by more than 11%. In contrast, the age composition effects deflated the age-independent death rate, l, throughout the period under reference. During 1950–1955, the age composition of the population deflated the age-independent death rate, l, by just 2%, whereas it deflated the age-independent death rate, l, by almost 50% during 2010–2015. As a result, the change in the age composition of the population inflated the intrinsic growth rate of the country throughout the period under reference. In other words, the natural population growth rate in the country has always been faster than the intrinsic growth rate throughout the period under reference.

2.4 Decomposition of the Change in Natural Population Growth Rate

The natural population growth rate in the country decreased from around 1.68% per year during 1950–1955 to around 1.27% per year during 2010–2015 (Table 2.3) or a decrease of around 25%. The period 1950–2015 can be divided into the period before 1980–1985 and the period after 1980–1985. During the period 1950–1985, the natural population growth rate in the country increased, instead decreased, to reach an all-time high of almost 2.3% per year or an increase of almost 36% during 1950–1985. Both the birth rate and the death rate decreased during this period but the decrease in the birth rate was slower than the decrease in the death rate. After 1980–1985, the rate of decrease in the birth rate accelerated, whereas the rate of decrease in the death rate decelerated so that the natural population growth rate started decreasing. During 1950–1985, the birth rate decreased by less than 19% whereas the death rate decreased by almost 53%. During 1980–2015, the birth rate decreased by almost 44%, whereas the death rate decreased by only about 42% so that the natural population growth rate decreased by more than 44%. Had the pace of reduction in the death rate would have been the same during 1980–2015 as during 1950–1985, the decrease in the natural population growth rate in the country would have been significantly slow (Fig. 2.4).

It can also be seen from Table 2.3 that the change in the intrinsic component tended to decrease the natural population growth rate but the age component tended

Table 2.3 Decomposition of the change in natural population growth rate in India

Period	Change in			Change in *r* attributed to the change in		Change in *b* attributed to the change in		Chang in *d* attributed to the change in	
	∇r	∇b	∇d	∇i	∇a	∇f	∇ab	∇l	∇ad
1950–1955/1955–1960	1.974	−0.980	−2.955	0.859	1.115	−1.154	0.173	−2.013	−0.942
1955–1960/1960–1965	1.599	−1.137	−2.736	0.415	1.184	−1.261	0.124	−1.676	−1.060
1960–1965/1965–1970	0.987	−1.647	−2.634	−0.039	1.026	−1.504	−0.142	−1.466	−1.168
1965–1970/1970–1975	0.976	−1.358	−2.334	−0.539	1.514	−1.759	0.400	−1.220	−1.114
1970–1975/1975–1980	0.285	−1.733	−2.018	−1.692	1.977	−2.626	0.893	−0.934	−1.084
1975–1980/1980–1985	0.274	−1.089	−1.362	−1.273	1.547	−1.869	0.781	−0.596	−0.766
1980–1985/1985–1990	−1.408	−2.528	−1.120	−2.562	1.154	−2.954	0.426	−0.392	−0.727
1985–1990/1990–1995	−1.575	−2.990	−1.415	−2.468	0.893	−2.930	−0.060	−0.461	−0.953
1990–1995/1995–2000	−1.357	−2.465	−1.108	−1.734	0.376	−2.115	−0.350	−0.381	−0.727
1995–2000/2000–2005	−1.630	−2.328	−0.698	−1.948	0.318	−2.224	−0.103	−0.277	−0.421
2000–2005/2005–2010	−1.853	−2.449	−0.596	−2.301	0.448	−2.555	0.106	−0.254	−0.342
2005–2010/2010–2015	−2.447	−2.931	−0.484	−2.531	0.084	−2.759	−0.172	−0.228	−0.256
1950–1955/2010–2015	−4.175	−23.635	−19.460	−16.003	11.828	−25.235	1.600	−9.232	−10.228

Source Author's calculations

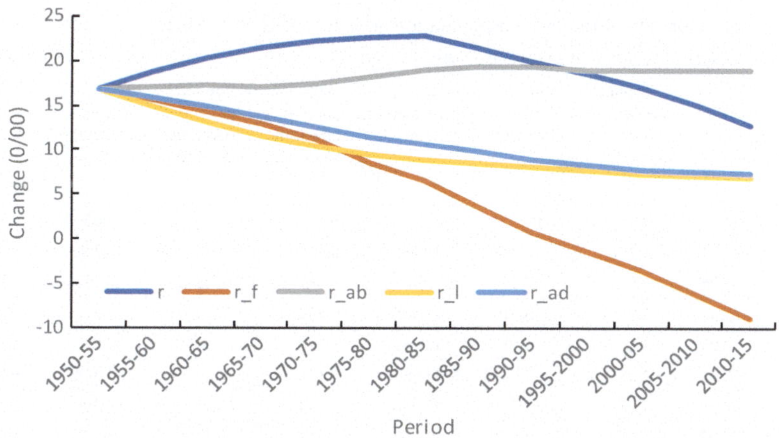

Fig. 2.4 Decomposition of the change in natural population growth rate in India, 1990–2015

to increase the natural population growth rate. The age component of the natural population growth rate increased throughout the period under reference. The magnitude of the decrease in the intrinsic component was smaller than that of the increase in the age component during 1950–1985. During 1980–2015, the decrease in the intrinsic component accelerated so that the natural population growth rate decreased after 1980–1985. On the other hand, age structure effects on the birth rate always contributed to decelerate the decrease in the birth rate relative to the decrease in the age-independent birth rate but always contributed to accelerate the decrease in the death rate relative to the decrease in the age-independent death rate.

2.5 Decomposition of the Change in Population Size

The net addition to the population of the country during 1950–1955 was around 33 million (Table 2.4). At this constant increase, India's population would have increased by almost 430 million to reach around 806 million in 2015 against the actual increase of almost 1315 million or an additional increase of around 509 million. The reason is that the five factors that determine the increase in population changed during this period. The fertility, measured in terms of the age-independent birth rate, decreased during this period, which resulted in a decrease of 506 million in the population between 1950 and 2015. The mortality, measured in terms of the age-independent birth rate, also decreased during this period, which resulted in an increase of 231 million in the population. Similarly, the change in the age structure effects on the birth rate resulted in an increase of 55 million, whereas the change in the age structure effects on the death rate resulted in an increase of 216 million in the population of the country. Finally, the increase in the size of the population between 1950 and 2010

Table 2.4 Components of increase in population size in India 1950–2015

Year	Constant increase	Increase attributed to the change in						Population (million)
		Population size	Age-independent birth rate	Age-independent death rate	Age composition effects on birth rate	Age composition effects on death rate	Total increase during the period	
1950								376.325
1955	33.05	0	0	0	0	0	33.050	409.375
1960	33.05	3.261	−2.370	4.135	0.356	1.934	40.366	449.740
1965	33.05	7.595	−5.214	7.916	0.635	4.326	48.307	498.047
1970	33.05	13.044	−8.970	11.575	0.280	7.243	56.221	554.269
1975	33.05	19.692	−13.857	14.965	1.391	10.34	65.580	619.849
1980	33.05	27.553	−22.013	17.865	4.164	13.705	74.325	694.173
1985	33.05	36.576	−28.513	19.938	6.879	16.370	84.300	778.473
1990	33.05	46.177	−40.010	21.466	8.539	19.201	88.422	866.895
1995	33.05	55.481	−52.701	23.463	8.279	23.330	90.902	957.797
2000	33.05	64.360	−62.824	25.287	6.603	26.808	93.285	1051.082
2005	33.05	72.650	−74.502	26.740	6.061	29.020	93.019	1144.101
2010	33.05	79.991	−89.096	28.191	6.667	30.974	89.776	1233.877
2015	33.05	85.923	−106.075	29.596	5.611	32.548	80.652	1314.528
1950–15	429.644	512.302	−506.145	231.136	55.467	215.799	938.203	

Source Author's calculations

resulted in an increase of 512 million in the population of the country (Fig. 2.5). The increase in population as a result of the effect of the change in the age structure of the population on the birth rate and the death rate and the increase in population as a result of the increase in population size are essentially the momentum effects.

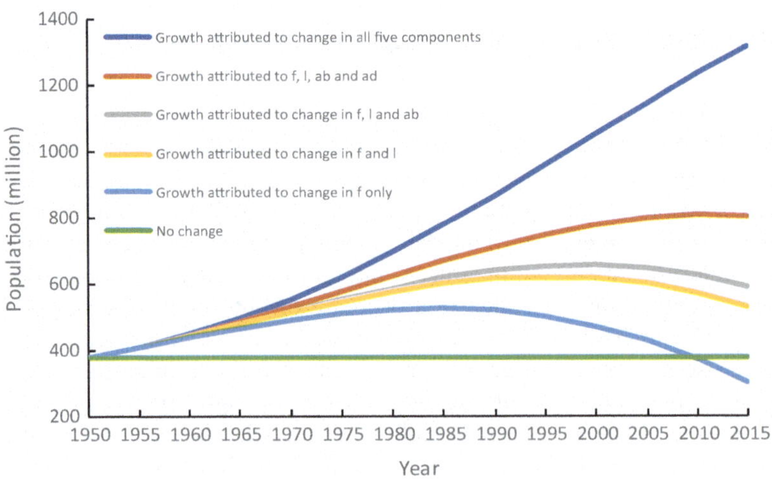

Fig. 2.5 Decomposition of population growth in India, 1990–2015

They are the result of the past fertility and mortality dynamics and are immune to interventions directed towards reducing fertility and mortality. This means that the momentum built in the population stock—size and the age composition—of the country is estimated to have resulted in an increase of around 784 million in India's population between 1950 and 2015. By comparison, the decrease in fertility and mortality during this period accounted for a decrease of about 275 million in the population. As a result, the actual increase in the population was almost 509 million more than the increase in the population that would have been there had the average annual addition to the population would have remained unchanged since 1950–1955. Obviously, the pace of the decrease in fertility and mortality in the country during the 65 years between 1950 and 2015 has not been fast enough to compensate for the increase in the population of the country attributed to the momentum built in the population stock of the country.

Alternatively, following Bongaarts and Bulatao (1999), India's population has multiplied by more than 3.49 times between 1950 and 2015. The increase in population size during 1950–2015 multiplied the population in 1950 by around 1.64 times; change in the age structure effects on the death rate multiplied the population by around 1.37 times; change in the age structure effects on the birth rate multiplied the population by around 1.10 times; decrease in mortality multiplied the population by around 1.77 times; and, finally, the decrease in fertility multiplied the population by around 0.80 times. In other words, the momentum for increase built in the population stock of the country multiplied the population in 1950 by around 2.48 times, whereas decrease in fertility and mortality multiplied the population in 1950 by around 1.41 times. The primary driver for controlling and curbing the increase in the population is the reduction in fertility. Table 2.4 suggests that the decrease in the population as a result of a decrease in fertility in the country during the 65 years between 1950 and 2015 had not been large enough to compensate for the increase in population attributed to the decrease in mortality and to the momentum for increase built in the population stock. Population transition has also resulted in significant changes in the age composition of the population of the country. The working age (15–64 years) population was around 59% of the total population of the country in 1950 (Table 2.5). This proportion decreased to an all-time low of around 55% in 1965 but increased consistently thereafter so that, by 2015, the working age population constituted almost two-thirds of the population of the country. More specifically, the ratio of the working age to non-working age population in the country decreased during 1950–1965 to reach an all-time low in 1965 when there were less than 125 working age (15–64 years) persons for every 100 non-working age (0–14 years and 65 years and above) persons in the country. However, this ratio increased consistently after 1965 so that, by 2015, there were almost 192 working age persons for every 100 non-working age persons in the country, indicating a substantial increase in the concentration of the population of the country in the working ages. This increase in the concentration of the population in the working ages is an indicator of the differential rate of growth of the working age population and the non-working age population. During the period 1950–1965, the growth of the working age population in the country was slower than that of the non-working age population so that the concentration

Table 2.5 Age structure transition in India 1950–2015

Year	Working age population (million)	Increase in working age population (million)										Ratio of working age to non-working age population	Difference in the growth rate of working and non-working age population
		15–19	20–24	25–29	30–34	35–39	40–44	45–49	50–54	55–59	60–64		
1950	223.427											1.461	0
1955	237.746	2.071	2.191	2.060	1.526	1.509	0.928	2.089	0.261	2.178	−0.495	1.386	0.740
1960	254.568	1.043	2.304	2.409	2.239	1.690	1.630	1.055	2.053	0.434	1.964	1.306	0.729
1965	275.493	4.540	1.410	2.628	2.652	2.422	1.852	1.735	1.162	1.993	0.530	1.240	0.775
1970	309.079	11.79	4.694	1.663	2.861	2.882	2.644	2.060	1.870	1.265	1.857	1.264	1.081
1975	350.665	6.926	12.15	5.209	2.207	3.301	3.246	2.946	2.300	1.986	1.314	1.296	1.094
1980	398.060	7.364	6.805	11.912	5.168	2.304	3.382	3.318	2.979	2.281	1.882	1.333	1.106
1985	449.823	7.105	7.266	6.710	11.741	5.135	2.338	3.363	3.239	2.817	2.048	1.356	1.063
1990	506.787	8.538	6.985	7.129	6.601	11.536	5.052	2.311	3.244	3.042	2.527	1.395	1.112
1995	569.664	9.053	8.466	6.903	7.047	6.538	11.354	5.001	2.347	3.205	2.962	1.458	1.184
2000	640.951	11.819	9.051	8.435	6.867	6.992	6.484	11.133	4.940	2.400	3.165	1.555	1.281
2005	714.451	7.044	11.602	8.822	8.232	6.703	6.837	6.343	10.769	4.781	2.365	1.663	1.309
2010	787.765	3.618	6.950	11.434	8.705	8.116	6.602	6.723	6.233	10.346	4.587	1.777	1.335
2015	860.128	3.959	3.901	7.108	11.427	8.688	8.084	6.571	6.675	6.166	9.784	1.916	1.429

Source Author's calculations

of the population in working ages decreased, instead increased. However, the growth rate of the working age population became faster than that of the non-working age population since 1965–1970 so that the concentration of the population of the country in working ages increased. This increase in the concentration of the population rendered a demographic advantage for economic growth in the country. However, the rate of increase in the concentration of the population in the working ages has at best been slow so that the potential contribution of the demographic transition to the economic growth of the country or the demographic dividend has never been large. In other words, the demographic dividend resulting from the transition in the age composition of the population in the country has never been able to have a telling impact on the economic growth. Although the difference between the growth rate of the working age population and the growth rate of the non-working age population in the country continues to increase, yet the net addition to the working age population of the country peaked during 2000–2005 and then decreased, which indicates that the expansion of the working age population of the country has started slowing down. This slowing down of the increase in the working age population has implications about the impact of population transition on the economic growth of the country.

2.6 Discussions and Conclusions

Population transition is implicit in the process of social and economic development. The experience of the European countries suggests that as social and economic development advances, death rate decreases and is followed, usually after a time lag, by the decrease in the birth rate. The natural population growth rate, therefore, first increases and then decreases. The time lag between the onset of the decrease in the death rate and the onset of the decrease in the birth rate determines the increase in population size and changes in population age composition which have economic implications. Although it is difficult to ascertain this 'normal' pace of transition, yet, the pace of population transition can be accelerated through explicit birth rate reduction policies as social and economic development induces a decrease in the death rate.

In line with the above arguments, India made specific efforts to accelerate the pace of population transition, although population transition in the country has been slow. India was the first country in the world to adopt an official population policy way back in 1952 and to launch an official programme to bring down the birth rate in an attempt to curb population growth. However, the natural population growth rate in the country started decreasing only after 1985 when the decrease in fertility accelerated and the decrease in mortality decelerated. Chaurasia (2010) has also observed that mortality transition in India slowed down particularly after 1980. The slow mortality transition in India may also be judged from the fact that between 1990 and 2015, the under-five mortality rate in the country decreased by only 62% (UNICEF 2015). Had the pace of mortality transition not decelerated after 1980, population transition in the country would have been even slower. The slow mortality transition also appears to be the reason behind the slow fertility transition in the country. It appears that the decrease in mortality in India has never been substantive enough to act as a catalyst for accelerating the decrease in fertility. The most commonly used indicator of fertility transition is the achievement of the replacement fertility. However, Espenshade et al. (2003) have shown that major determinant of cross-national variation in replacement fertility is the difference in mortality. In countries where mortality is high, replacement fertility is much higher than the widely assumed global average of 2.1 life time births per woman. This means that the proportion of young population remains large and hence the demographic dividend small.

Accelerating mortality transition to accelerate the pace of population transition in India remains a major public health challenge to India even today.

Population transition in India also does not appear to be sensitive to the population and development policies of the country adopted from time to time, although commitment to reducing fertility and curtailing population growth has explicitly been stated in all Five-year Development Plans and population and health policies of the country that were launched since independence. The country has always been under-achiever in terms of demographic goals and targets set at the policy level. For example, the target of achieving a crude birth rate of 25 live births per 1000 population by the year 1972 was set up way back in 1962 but could be achieved 30 years later—in 2002 only (Chaurasia and Gulati 2008). Similarly, the target of achieving replacement fertility by 2010 in the National Population Policy 2000 is yet to be achieved. The observed insensitiveness of population transition to official population and development policies appears to be a major reason behind the slow population transition in the country.

The analysis presented in this chapter also suggests that the momentum for growth built in the population stock—size and age composition—has increasingly become important in deciding population growth in the country. The relative importance of the age component in deciding the natural population growth rate has increased from less than 20% during 1950–1955 to almost 80% during 2010–2015. Obviously, population growth in the country is now increasingly being driven by the momentum built in the population stock rather than by the prevailing levels of fertility and mortality. This observation has important implications for population and development policies. The age component of the natural population growth rate reflects the age composition effects on birth and death rates which are essentially arithmetic accounting effects. They do not affect the level of fertility and mortality but influence birth and death rates. They are the result of the levels of fertility and mortality in the past. This observation is important from the policy perspective as the age composition effects on the birth rate and the death rate cannot be changed. They can be addressed by increasing the age at marriage and extending the interval between successive births (Andreev et al. 2013; Bongaarts 1994).

India requires appropriate policy response to the current state of population transition so that the change in the population stock in the coming years remains in tune with sustainable development processes. However, the country remains largely insensitive to the current state of population transition even at the policy level. It continues to persist with the National Population Policy 2000 despite the fact that policy goal of achieving the replacement fertility by 2010 could not be achieved and the medium-term goal of stabilising population growth by 2040 appears only a distant possibility. The total fertility rate in India still continues to hover around 2.2–2.3 life time births per woman of reproductive age (Government of India 2016) but the fertility of married women remains very high—4.6 life time births per married women of reproductive age (Government of India 2016). At the same time, the expectation of life at birth remains well below 70 years (Government of India 2016). Recently, the Government of India has announced a new health policy which aims at reducing the total fertility rate to the replacement level and increasing the expectation of life at birth to 70 years by the year 2025 (Government of India 2017), although population stabilisation has been identified as one of the thrust areas. On the other hand, in reply to a question in India's Parliament, the Health Minister of India has argued that the National Population Policy 2000 was uniformly applicable to all irrespective of religions and communities, and there was no proposal under consideration of the Government to formulate new uniform population policy (Government of India 2015). It is, however, imperative that priorities for addressing population trends in the country should partly be informed by the relative contribution of different demographic drivers of population change. Since the relative contribution of different demographic drivers on population growth is different at different stages of population transition, the formulation and adoption of population policy should be a dynamic entity that should keep changing as the population transition advances. This has, unfortunately, not been the case in India.

References

Adlakha A, Banister J (1995) Demographic perspectives on China and India. J Biosoc Sci 27(2):163–178

Andreev K, Kantorova V, Bongaarts J (2013) Demographic components of future population growth. New York, United Nations, Population Division. Technical Paper No. 2013/3

Bloom DE (2011) Population dynamics in India and implications for economic growth. Harvard University, Program on the Global Demography of Ageing. PGDA Working Paper No. 65

Bloom DE, Williamson JG (1998) Demographic transitions and economic miracles in emerging Asia. World Bank Econ Rev 12(3):419–456

Bloom DE, Canning D, Malaney PN (2000) Population dynamics and economic growth in Asia. Popul Dev Rev 26(Supplement) Population and Economic Change in East Asia: 257–290

Bloom DE, Canning D, Sevilla J (2002) The demographic dividend: a new perspective on economic consequences of population change. Santa Monica, CA, RAND, MR-1274

Bloom DE, Canning D (2004) Global demographic change: dimensions and economic significance. National Bureau of Economic Research. Working Paper No. 10817

Bloom DE, Canning D (2005) Global demographic change: dimensions and economic significance. In: Global demographic change: economic impacts and policy challenges. Federal Reserve Bank of Kansas City

Bloom DE, Canning D, Wu L, Liu Y, Mahal A, Yip W (2006) Demographic change and economic growth: comparing China and India. Harvard University, Harvard School of Public Health, Boson

Bongaarts J (1994) Population policy options in the developing world. Science 263:2–7

Bongaarts J (2009) Human population growth and the demographic transition. Philos Trans R Soc B Biol Sci 364(1532):2985–2990
Bongaarts J, Bulatao RA (1999) Completing the demographic transition. Popul Dev Rev 25(3):515–529
Bourgeois-Pichat J, Taleb SA (1970) Un taux d'accroissement nul pour les pays en voie dédeveloppement en l'an 2000. Réve or réalité? Population 25(5):951–974
Cai F, Wang D (2005) Demographic transition: implications for growth. In: Garnaut R and Song L (eds) The China boom and its discontents. Asia-Pacific Press, Canberra
Chaurasia AR (2010) Mortality transition in India: 1970-2005. Asian Population Studies 6(1):47–68
Chaurasia AR (2017) Fertility, mortality and age composition effects of population transition in China and India. Comp. Population Stud. 42:149–186
Chaurasia AR, Gulati SC (2008) India: the state of population 2007. National Population Commission and Oxford University Press, New Delhi
Chesnais J (1979) L'effet multiplicatif de la transition démographique. Population 36(6):1138–1144
Chesnais J (1986) La transition démographique: Etapes, formes, implications économiques. Cahier INED No. 113. Presses Universitaires de France, Paris
Chesnais J (1990) Demographic transition patterns and their impact on the age structure. Popul Dev Rev 16(2):327–336
Choudhry MT, Elhorst JP (2010) Demographic transition and economic growth in China, India and Pakistan. Econ Syst 34:218–236
Coale AJ (1983) Population trends in China and India (A review). Proc Natl Acad Sci USA 80:1757–1763
Da Vanzo J, Dogo H, Grammich CA (2011) Demographic trends, policy influences, and economic effects in China and India through 2025. Santa Monica, RAND Corporation, National Security Research Division. Working Paper WR-849
Desai, M (2003) India and China: an essay in comparative political economy. Paper presented at IMF Conference A Tale of Two Giants: India's and China's Experience with Reform and Growth. New Delhi, National Council of Applied Economic Research
Espenshade TJ (1975) The stable decomposition of the rate of natural increase. Theor Popul Biol 8:97–115
Espenshade TJ, Guzman JC, Westoff CF (2003) The surprising global variation in replacement fertility. Popul Res Policy Rev 22:575–583
Espenshade TJ, Tannen JBC (2015) Population dynamics. Momentum of population growth. In: Wright JD (ed) International encyclopedia of social and behavioral sciences, 2nd edn, vol 18
Frejka T (1973) The future of population growth: alternative paths to equilibrium. Wiley, New York
Frejka T (1981) Long-term prospects for world population growth. Popul Dev Rev 7(3):489–511
Golley J, Tyers R (2012) Demographic dividends, dependencies and economic growth in China and India. University of Western Australia Business School. Discussion Paper 12.03
Government of India (2000) National population policy 2000. Ministry of Health and Family Welfare, New Delhi
Government of India (2014) Census of India 2011. Report on post enumeration survey. Registrar General and Census Commissioner of Infia, New Delhi
Government of India (2015) National population policy. Ministry of Health and Family Welfare. Press Information Bureau, New Delhi
Government of India (2016) Sample Registration System statistical report 2015. Registrar General and Census Commissioner of India, New Delhi
Government of India (2017) National health policy 2017. Ministry of Health and Family Welfare, New Delhi
Horiuchi (1991) Measurement and analysis of cohort size variations. Population Bulletin of the United Nations 30: 106–124
Horiuchi S (1992) Stagnation in the decline of the world population growth rate during the 1980s. Science 257:761–765

Horiuchi S (1993) The world population growth rate: why declines stalled in the 1980s. Popul Today 21(6):6–9

Horiuchi S (1995) The cohort approach to population growth: a retrospective decomposition of growth rates for Sweden. Population Studies 49(1995):147–163

Joe W, Dash AK, Agrawal P (2015) Demographic transition, savings and economic growth in China and India. Delhi, Institute of Economic Growth. Working Paper No. 351

Kandiah V, Horiuchi S (1995) Recent trends and prospects in world population growth. Popul Bull UN 39:133–146

Kelkar V (2004) India: on the growth turnpike. Narayanan oration speech. Australian National University, Canberra

Keyfitz N (1971) On the momentum of population growth. Demography 8(1):71–80

Kitagawa EM (1955) Components of difference between two rates. J Am Stat Assoc 50(272):1168–1194

Mason A (2007) Demographic transitions and demographic dividends in developed and developing countries. In: United Nations expert group meeting on social and economic implications of changing population age structures. United Nations, Department of Economic and Social Affairs, Population Division, New York

Notestein FW (1945) Population: the long view. In: Schultz T (ed) Food for the World. University of Chicago Press, Chicago, pp 36–57

Notestein FW (1960) Mortality, fertility, the size-age distribution and the growth rate. In: National bureau of economic research, demographic and economic change in developed countries. Columbia University Press

Preston SH, Guillot M (1997) Population dynamics in an age of declining fertility. Genus 53(3–4):15–31

Preston SH, Heuveline P, Guillot M (2001) Demography. Measuring and modeling population processes. Blackwell Publishers, Oxford, p 136

Roy SG (1993) Demography of China and India: a comparative study. In: International Union for the Scientific Study of Population (IUSSP) international population conference. Liege, Belgium, IUSSP:173–186

Schmidt JD (2011) India and China in comparative perspective-emerging Asian and global powers. Paper presented at Madras Institute of Development Economics, Chennai

Schoen R (2002) Mathematical models in demography and actuarial mathematics. University Park, Pennsylvania State University, Population Research Institute. Working Paper No. 02–07

Schoen R, Kim YJ (1991) Convergence towards stability is a fundamental principle of population dynamics. Demography 28(3):455–466

Smith SK, Ahmed B (1990) A demographic analysis of the population growth of the states, 1950-1980. J Reg Sci 30(2):209–227

United Nations (2006) World Population Prospects: The 2004 Revision. Volume III. Analytical Report. New York, United Nations, Department of Economic and Social Affairs. Population Division

United Nations (2017) World Population Prospects: The 2017 Revision. DVD Edition. New York, United Nations, Department of Economic and Social Affairs. Population Division

United Nations Children's Fund (2015) Levels and Trends in Child Mortality. Repot 2015. Estimates Developed by United Nations Inter-Agency Group for Child Mortality Estimation. New York, United Nations Children's Fund

Vallin J (2006) Europe's demographic transition 1740-1940. In: Caselli G, Vallin J, Wunsch G (eds) Demography: analysis and synthesis. New York, Academic Press

White MJ, Subedi I (2008) The demography of China and India: effects on migration to high income countries through 2030. Migration Policy Institute, Washington DC

Williamson J (1977) Growth, distribution and demography: some lessons from history. National Bureau of Economic Research. Working Paper No, 6244, Cambridge, MA

Wolf C Jr, Dalal S, DaVanzo J, Larsen EV, Akhmedjonov A, Dogo H, Huang M, Montoya S (2011) China and India, 2025. A Comparative Assessment. RAND Corporation, Santa Monica

Chapter 3
Future Population Growth, 2015–2100

Abstract This chapter analyses the prospects of population growth in India for the period 2015–2100 on the basis of the projections prepared by the United Nations Population Division. The chapter decomposes the projected population growth in the country during 2015 through 2100 into the projected population growth attributed to the change in fertility and mortality, to the change in the age and sex structure effects on the birth rate and the death rate and to the change in the size of the population to identify main drivers of the future population growth. The analysis suggests that the projected increase in India's population will be driven by its very large population size as the transition in fertility and mortality and the change in population age and sex structure will contribute to decrease, instead increase, the population of the country.

3.1 Introduction

The population projections prepared by the Population Division of the United Nations in 2017 suggest that India's population is most likely to increase from an estimated about 1309 million in 2015 to around 1679 million by 2061 and then to decrease to about 1577 million by 2100 (United Nations 2017b). These estimates are based on the assumption that the country will achieve the replacement fertility sometimes during 2025–2030. It is also projected that the total fertility rate in the country will continue to decrease till 2070–2075 to reach 1.77 births per woman of reproductive age, will remain unchanged till 2095 and will increase marginally to reach 1.78 births per woman of reproductive age during 2095–2100. It is also assumed that the expectation of life at birth will increase consistently to more than 81 years during 2095–2100. Along with the increase in the size of the population, there will be significant changes in population age and sex composition leading to significant changes in the population stock of the country which will have implications to sustainable development. The projections prepared by the United Nations also indicate that India is most likely to become the most populous country in the world by 2022 surpassing the population of China (United Nations 2017b). The projection exercise carried out by the Population Division of the United Nations also suggests that if the country was able to achieve the replacement fertility by the year 2010 as envisioned in the National Population

© Springer Nature Singapore Pte Ltd. 2020

A. R. Chaurasia, *Population and Sustainable Development in India*,
https://doi.org/10.1007/978-981-32-9212-3_3

Policy 2000 (Government of India 2000), then the population of the country would have increased to almost 1486 million by the year 2042 and then decreased to around 907 million by the turn of the century. However, if, under the unlikely scenario that the country is not able to achieve the replacement fertility by the end of the present century, then the population of the country will soar up to more than 2388 million by the year 2100. The projected trend in the population of the country will be associated with significant changes in its population stock as it is well known that in a population closed to migration, the change in the population stock between two points in time is determined by the initial size of the population, changes in the levels of fertility and mortality and changes in the population age and sex structure. The contribution of the change in the initial population size and in the age and sex structure of the population is closely related to the concept of population momentum (Keyfitz 1971; Schoen and Kim 1991; Horiuchi 1995).

This chapter decomposes the projected population growth in India during 2015 through 2100 into the projected population growth attributed to the change in fertility and mortality, to the change in the age and sex structure effects on the birth rate and the death rate and to the change in the size of the population to identify main drivers of the population growth in the country during 2015–2100. The analysis follows the decomposition methodology used in the previous chapter to analyse the historical trend in the growth of the population of the country. Such an analysis has relevance to population and sustainable development policies and programmes. It is obvious that if the changes in the age structure of the population and in the initial population size account for, say, 80% of the future population growth, then there is little relevance of fertility and mortality reduction policies and programmes in the context of population stabilisation and the impact of these policies and programmes on sustainable development. In such a scenario, population influencing policies and programmes should be directed to addressing the momentum effects of future population growth which are the result of the fertility and mortality levels in the past. The momentum effects of population growth can at best delayed through appropriate policy and programme interventions. They cannot be eliminated. (Bongaarts 1994).

The analysis of the demographic drivers of future population growth has always been an area of interest to demographers. The approach that has commonly been adopted for such analyses is based on the projection of the population stock at a future date under a set of assumptions about future trends in fertility and mortality (Andreev 2013; Bongaarts 1994, 2009; Bongaarts and Bulatao 1999; Frejka 1973, 1981). These analyses are built upon the hypothetical cohort-component projection methodology which is based on the classical demographic transition model. For example, the recent paper by Andreev (2013) compares four types of population projections—standard, natural, replacement and momentum—to analyse the contribution of different demographic components to future world population growth. This analysis suggests that population momentum resulting from changes in the age and sex structure of the population will be the main demographic driver of the future population growth in the world under the assumption that fertility will sooner or later decrease to the replacement level and will remain at that level for a long period. As regards India, it has been estimated that the change in the age structure of the population, alone, will account for an increase of around 447 million in the population of the country between 2010 and 2100 (Chaurasia and Gulati 2008).

3.2 Data and Method

The analysis is based on the estimates and projections of population prepared by the United Nations Population Division in 2017 (United Nations 2017b). The preparation of each new revision of the world population prospects involves two distinct processes. The first is related to the incorporation of new information about the demography of each country or area of the world and, in some cases, a reassessment of the past. The second, on the other hand, is related to the formulation of detailed assumptions about the future path of fertility, mortality and international migration for every country or area of the world (United Nations 2017b). The projection exercise follows the most commonly used component projection method. This method involves calculation of the effect of the assumed future course of fertility, mortality, and international migration on population at some given point in the future (Preston et al. 2001). The underlying assumption is that both fertility and mortality will continue to decrease in future. Three assumptions have been made about the future course of the decrease in fertility—medium decrease, slow decrease and fast decrease—while only one assumption is made about the future course of the decrease in mortality. Based on the assumptions about the project decrease in fertility and mortality and projected trend in international migration, nine variants of population projections have been prepared for each country in the 2017 revision: (1) low fertility variant; (2) medium fertility variant; (3) high fertility variant; (4) constant fertility variant; (5) instant replacement fertility variant; (6) constant mortality variant; (7) no change variant; (8) zero-migration variant; and (9) momentum variant. Details of the assumptions regarding future trend in fertility, mortality and international migration related to different variants of population projections prepared by the United Nations Population Division are given elsewhere and not repeated here (United Nations 2017a). The medium variant of population projection is argued to be the most likely future scenario. In the present analysis, however, we have used the zero-migration variant of the latest population projections prepared by the United Nations Population Division. Assumptions related to this variant of population projection are the same as the medium variant of population projection, with the only difference that the international migration is set to zero throughout the projection period. This assumption essentially implies that the projected growth of the population is the result of the projected change in the annual number of births and annual number of deaths resulting from the change in the levels of fertility and mortality.

The Government of India also carries out an exercise to project the population of the country after every decennial population census. The projection horizon of this exercise is, however, limited to 25 years only. Moreover, the exercise carried out by the Government of India makes only one assumption about the future trend in fertility and mortality. The projection of the country's population based on the data available

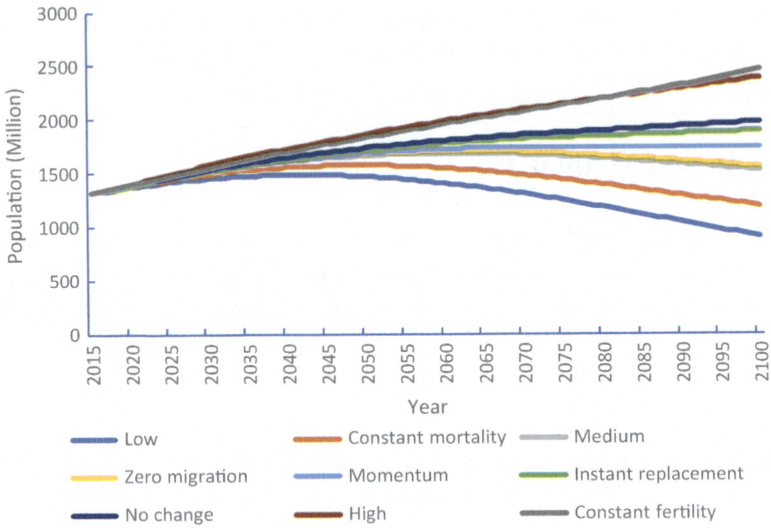

Fig. 3.1 Projected population growth in India, 2015–2100

through 2001 population census is currently available (Government of India 2006). Population projection based on the data available through 2011 population census is yet to be released by the Government of India. A comparison of the projected population carried out on the basis of 2001 population census by the Government of India for the year 2011 with the population enumerated at the 2011 population census has revealed that the population enumerated at the 2011 population census was more than 18 million larger than the population projected on the basis of 2001 population census. At the same time, the post-enumeration check conducted after the 2011 population census revealed an omission rate of around 2.3% (Government of India 2014) which means that the actual increase in the population of the country during 2001–2011 was more than 41 million higher than the projected one.

Figure 3.1 and Table 3.1 present the projected population of the country under nine variants of the projection exercise carried out by the Population Division of the United Nations. The table suggests that country's population will be somewhere between 907 and 2453 million by the turn of the century depending upon the future course of fertility, mortality and migration. There are only two variants: low fertility variant and constant mortality variant, under which the population of the country in the year 2100 is projected to be lower than the population of the country estimated in the year 2015.

The methodology adopted for analysing the drivers of the future population growth is the same as the methodology adopted in the previous chapter for analysing the historical trend in population. This methodology, essentially, decomposes the projected growth in the population into the growth attributed to the projected change in fertility; projected change in mortality; associated change in the age structure effects on birth rate and death rates; and the projected growth attributed to the increase in the popu-

Table 3.1 Projected population of India (million) under different projection variants of United Nations Population Division

Year	Medium	High	Low	Zero-migration	Constant fertility	Instant replacement	Momentum	Constant mortality	No change
2015	1309.054	1309.054	1309.054	1309.054	1309.054	1309.054	1309.054	1309.054	1309.054
2020	1383.198	1396.360	1370.036	1385.829	1390.107	1380.892	1381.668	1379.814	1386.670
2025	1451.829	1486.794	1416.864	1456.938	1472.009	1451.727	1451.386	1442.020	1461.926
2030	1512.985	1575.831	1450.140	1520.770	1551.869	1518.957	1515.928	1493.981	1532.112
2035	1564.570	1655.526	1473.701	1575.164	1626.732	1579.713	1572.158	1533.661	1594.243
2040	1605.356	1727.110	1485.033	1618.865	1696.157	1631.702	1617.935	1560.020	1647.881
2045	1636.496	1795.702	1483.026	1653.035	1763.164	1676.172	1654.354	1573.862	1695.420
2050	1658.978	1863.473	1468.059	1678.666	1828.999	1715.302	1683.948	1576.385	1738.114
2055	1673.078	1928.776	1441.656	1695.898	1893.094	1749.993	1707.656	1568.202	1775.592
2060	1678.568	1988.739	1405.371	1704.445	1954.276	1779.606	1725.058	1549.641	1807.096
2065	1675.744	2042.955	1360.374	1704.558	2012.806	1803.481	1735.779	1521.530	1833.141
2070	1665.179	2093.121	1307.207	1696.778	2070.107	1821.770	1740.648	1485.215	1855.541
2075	1648.425	2142.038	1247.167	1682.642	2128.090	1836.187	1741.624	1442.565	1876.176
2080	1626.927	2191.156	1181.846	1663.583	2187.635	1848.444	1740.832	1395.572	1896.160
2085	1602.017	2240.587	1113.242	1640.927	2249.154	1859.788	1739.662	1345.988	1915.947
2090	1574.797	2289.869	1043.450	1615.766	2313.406	1871.030	1738.585	1295.027	1935.855
2095	1545.959	2338.807	974.063	1588.791	2381.006	1882.285	1737.564	1243.715	1956.256
2100	1516.597	2388.350	906.865	1561.104	2452.765	1894.127	1736.525	1192.910	1977.340

Source United Nations (2017a)

lation stock during 2015–2100. The components of the projected population growth are not the same for different variants of population projections as the assumptions related to projected fertility, mortality and international migration are different for different variants of population projection. The present analysis is, therefore, limited to the zero-migration variant of United Nations population projection exercise.

3.3 Future Population Growth

The medium variant of the population projections prepared by the United Nations in 2017 suggests that India's population will continue to increase till 2061 to peak around 1679 million and then decrease to around 1517 million by the turn of the century. This means that around 370 million people will be added to the population of the country during 2015–2061 and then country's population will decrease by around 162 million during 2061–2100 so that the net addition to the population of the country during 2015–2100 will be around 208 million. This medium variant projection is based on the assumption that the total fertility rate will reach the replacement level some time during 2025–2030 and will continue to decrease till 2070–2075 to reach an all-time low of 1.77 births per woman of reproductive age. It is also projected that the total fertility rate will remain unchanged at this level till 2095 but will increase marginally to 1.78 births per woman of reproductive age during 2095–2100. At the same time, it is projected that the expectation of life at birth will increase steadily from less than 69 years during 2015–2020 to more than 81 years during 2095–2100. Moreover, it is projected that around 29 million people will migrate out of the country during 2015–2100 (Fig. 3.2).

However, if it is assumed that there will be no migration out of the country during the period under reference, then the zero-migration variant of United Nations population projections suggests that the population of the country will increase to around 1706 million by 2063 and then decrease to around 1561 million by 2100 so that

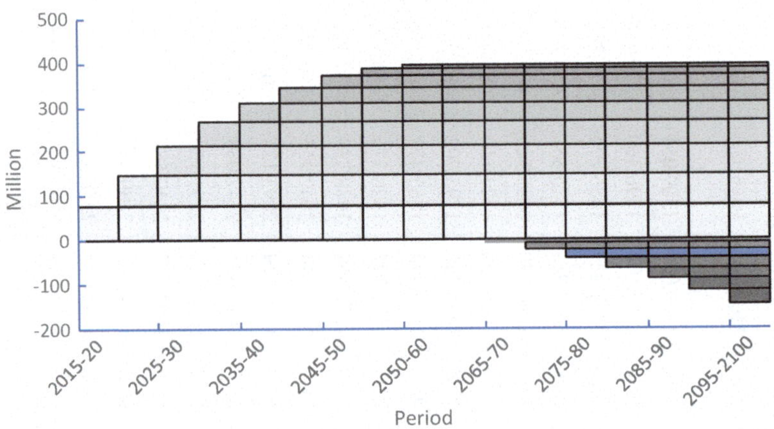

Fig. 3.2 Projected net addition to India's population in different quinquennials of the period 2015–2100

the net addition to country's population during 2015–2100 will be around 252 million. Key indicators of the future population of the country based on the zero-migration variant of population projection are presented in Table 3.2. The birth and death rates are projected to decrease throughout the current century so that the average annual natural population growth rate in the country will continue to decrease throughout the period under reference. However, the average annual natural population growth rate will turn negative only during 2065–2070. Moreover, the projected trend in the total fertility rate and the expectation of life at birth during 2015–2100 will be very similar under the zero-migration variant of population projection and under the medium variant version of the population projection exercise carried out by the United Nations. Lastly, the number of women in the reproductive age group (15–49 years) under the zero-migration version is projected to increase to more than 347 million by the year 2045. After 2045, the number of women in the reproductive age group is projected to decrease till 2100 so as to reach around 303 million by the end of the current century.

Table 3.2 Projected population, birth rate, death rate, total fertility rate, expectation of life at birth and women aged 15-49 years in India 2015–2100 under zero-migration variant of 2017 revision of UN population projections *Zero-migration variant of 2017 revision of UN population projections*

Year	Population (million)	Birth rate (Per 1000 popul ation)	Death rate (Per 1000 popul ation)	Total fertility rate	Expectation of life at birth (years)	Women (15–49 years) (Million)
2015	1309.054					336.722
2020	1385.829	18.76	7.36	2.30	68.89	358.982
2025	1456.938	17.53	7.52	2.19	70.03	377.319
2030	1520.770	16.37	7.79	2.10	71.00	388.515
2035	1575.164	15.16	8.13	2.02	71.90	394.604
2040	1618.865	14.03	8.56	1.95	72.70	397.068
2045	1653.035	13.18	9.00	1.90	73.48	397.139
2050	1678.666	12.57	9.49	1.86	74.22	393.680
2055	1695.898	12.05	10.01	1.83	74.96	386.860
2060	1704.445	11.57	10.56	1.81	75.68	378.307
2065	1704.558	11.13	11.12	1.79	76.40	371.236
2070	1696.778	10.76	11.68	1.78	77.10	362.635
2075	1682.642	10.48	12.16	1.77	77.82	352.804
2080	1663.583	10.28	12.56	1.77	78.54	342.141
2085	1640.927	10.13	12.87	1.77	79.26	331.555
2090	1615.766	9.98	13.07	1.77	80.00	321.654
2095	1588.791	9.84	13.21	1.77	80.72	312.240
2100	1561.104	9.73	13.24	1.78	81.47	303.052

Source United Nations (2017a)

The projected increase in the population of the country during 2015–2100 suggests that the natural population growth rate in the country will decrease throughout the current century. However, it will turn negative only during 2065–2070 and will remain negative throughout the period 2065–2100. Moreover, the birth rate independent of age structure effects, *f*, and the death rate independent of age structure effects, *l*, will also continue to decrease throughout the period under reference and the decrease in *f* will be faster than that in *l* so that the intrinsic rate of growth will turn negative from 2040–2045 onwards. This means that the positive natural population growth rate during 2040–2065 will be because of the age component of the natural population growth rate.

Table 3.3 also suggests that the age structure effects will continue to inflate the age-independent birth rate, *f*, till 2035–2040. After 2035–2040, these effects will contribute to deflate the age-independent birth rate so that the observed birth rate will be lower than the birth rate determined by the prevailing level of fertility. On

Table 3.3 Projected trend in natural population growth rate, fertility, mortality and age and sex structure effects on the birth rate and the death rate in India from 2015 to 2100

Period	*r*, Natural population growth rate (0/00)	Components of natural population growth rate		*f*, Birth rate independent of age structure effects (0/00)	*l*, Death rate independent of age structure effects (0/00)	*ab*, Age structure effects on the birth rate (0/00)	*ad*, Age structure effects on the death rate (0/00)
		i, Intrinsic component (0/00)	*a*, Age component (0/00)				
2015–2020	11.400	2.479	8.922	16.994	14.515	1.104	0.507
2020–2025	10.009	1.939	8.070	16.219	14.280	1.081	0.526
2025–2030	8.577	1.329	7.248	15.413	14.084	1.062	0.553
2030–2035	7.029	0.685	6.344	14.594	13.909	1.039	0.585
2035–2040	5.473	0.083	5.390	13.839	13.756	1.014	0.622
2040–2045	4.178	−0.432	4.609	13.178	13.610	1.000	0.661
2045–2050	3.077	−0.858	3.936	12.614	13.473	0.996	0.704
2050–2055	2.043	−1.253	3.295	12.087	13.340	0.997	0.750
2055–2060	1.005	−1.609	2.615	11.605	13.214	0.997	0.799
2060–2065	0.013	−1.844	1.857	11.245	13.089	0.990	0.849
2065–2070	−0.915	−1.997	1.082	10.973	12.969	0.981	0.900
2070–2075	−1.673	−2.126	0.452	10.724	12.850	0.977	0.946
2075–2080	−2.278	−2.236	−0.043	10.497	12.733	0.979	0.986
2080–2085	−2.743	−2.306	−0.436	10.310	12.617	0.982	1.020
2085–2090	−3.090	−2.356	−0.734	10.144	12.500	0.984	1.045
2090–2095	−3.367	−2.379	−0.988	10.010	12.389	0.983	1.066
2095–2100	−3.516	−2.362	−1.154	9.912	12.274	0.981	1.079

Source Author's calculations

the other hand, the age structure effects on the death rate are projected to increase throughout the current century. These effects will deflate the age-independent death rate, *l*, till 2075–2080 and then will start inflating the age-independent death rate, *l*. The decrease in the deflating effect of the population age structure on the age-independent death rate, *l*, is a reflection of the fact that the population of the country will be getting increasingly older with transition as a result of the decrease in fertility and mortality. The age component of the natural population growth rate will turn negative during 2075–2080 so that, after 2075–2080 onwards, both the intrinsic and the age components of the natural population growth rate will contribute to decreasing the natural population growth rate, thereby accelerating the decrease in the population of the country. It is projected that during 2095–2100, India's population will decrease at a rate of more than 3.5% per year.

3.4 Drivers of Future Population Growth

Table 3.4 decomposes the change in the natural population growth rate, *r*, into the change in the age-independent birth rate, *f*; age-independent death rate, *l*; age structure effects on the birth rate, *ab*; and age structure effects on the death rate, *ad*. The decrease in the age-independent birth rate, *f*, and age structure effects on birth rate, *ab*, decreases the natural population growth rate, *r*. On the other hand, the decrease in the age-independent death rate, *l*, and the age structure effects on the death rate, *ad*, increases the natural population growth rate, *r*. It may be seen from the table that the main drivers of the decrease in the natural population growth rate in India in the coming years will be the decrease in the age-independent birth rate and the increase in the age structure effects on the death rate. On the other hand, the decrease in the age-independent death rate will tend to marginally slowdown the decrease in the natural population growth rate.

Table 3.5 presents the projected change in the net addition to the population in different quinquennials of the period 2015 through 2100. During 2015–2020, the projected net addition to the population is projected to be around 77 million but, during 2020–2025, the net addition to the population is projected to be around 6 million less than the projected net addition to the population during 2015–2020. The net addition to the population is projected to decrease in every quinquennial till 2055–2060. During 2060–2065, there will be virtually no change in the population of the country. After 2060–2065, the net addition to the population is projected to turn negative so that, during 2065–2075, the population of the country is projected to decrease by around 8 million. The net decrease in the population in a quinquennial will then increase subsequently so that, during 2095–2100, the population of the country is projected to decrease by around 28 million.

As shown in the previous chapter, the change in the net increase or decrease in population in a quinquennial can be attributed to the change in: (i) the population size; (ii) the age-independent birth rate, *f*; (iii) the age-independent death rate, *l*; (iv) the age structure effects on the birth rate, *ab*; and (v) the age structure effects on the

Table 3.4 Decomposition of the change in the natural population growth rate in India from 2015 to 2100

Period	∇r 0/00	∂f 0/00	∂l 0/00	∂ab 0/00	∂ad 0/00	$\partial i = \partial f - \partial l$ 0/00	$\partial a = \partial ab - \partial ad$ 0/00
2015–2020/2020–2025	−1.392	−0.846	−0.121	−0.386	0.280	−0.725	−0.667
2020–2025/2025–2030	−1.432	−0.864	−0.106	−0.293	0.381	−0.758	−0.674
2025–2030/2030–2035	−1.548	−0.860	−0.100	−0.348	0.439	−0.761	−0.787
2030–2035/2035–2040	−1.556	−0.775	−0.092	−0.354	0.519	−0.682	−0.873
2035–2040/2040–2045	−1.296	−0.666	−0.094	−0.190	0.534	−0.572	−0.724
2040–2045/2045–2050	−1.100	−0.562	−0.093	−0.049	0.583	−0.469	−0.632
2045–2050/2050–2055	−1.035	−0.525	−0.096	0.011	0.617	−0.429	−0.606
2050–2055/2055–2060	−1.037	−0.481	−0.098	−0.004	0.650	−0.383	−0.654
2055–2060/2060–2065	−0.992	−0.358	−0.103	−0.078	0.660	−0.254	−0.738
2060–2065/2065–2070	−0.928	−0.268	−0.104	−0.100	0.664	−0.164	−0.765
2065–2070/2070–2075	−0.758	−0.243	−0.110	−0.036	0.589	−0.133	−0.625
2070–2075/2075–2080	−0.605	−0.223	−0.113	0.020	0.516	−0.109	−0.496
2075–2080/2080–2085	−0.464	−0.183	−0.116	0.030	0.428	−0.067	−0.398
2080–2085/2085–2090	−0.348	−0.164	−0.120	0.013	0.317	−0.043	−0.304
2085–2090/2090–2095	−0.277	−0.132	−0.118	−0.007	0.256	−0.014	−0.263
2090–2095/2095–2100	−0.149	−0.095	−0.123	−0.015	0.161	0.027	−0.176
2015–2020/2095–2100	*−14.916*	*−7.383*	*−1.777*	*−1.648*	*7.662*	*−5.606*	*−9.310*

Source Author's calculations

Table 3.5 Decomposition of the change in the net addition to the population in India from 2015 to 2100

Period	Change in the net addition to the population	Change in the net addition to the population attributed to						
		Change in population size	Change in age-independent birth rate	Change in age-independent death rate	Change in age structure effects on birth rate	Change in age structure effects on death rate	Change in the intrinsic component of growth rate	Change in age component of growth rate
2015–2020/2020–2025	−5.666	3.963	−5.855	−2.673	−0.839	1.941	−5.016	−4.613
2020–2025/2025–2030	−7.277	3.140	−6.282	−2.132	−0.772	2.774	−5.511	−4.906
2025–2030/2030–2035	−9.438	2.310	−6.530	−2.638	−0.756	3.336	−5.775	−5.974
2030–2035/2035–2040	−10.693	1.536	−6.092	−2.784	−0.727	4.081	−5.365	−6.864
2035–2040/2040–2045	−9.530	0.941	−5.381	−1.537	−0.758	4.312	−4.622	−5.848
2040–2045/2045–2050	−8.540	0.543	−4.641	−0.403	−0.771	4.810	−3.870	−5.213
2045–2050/2050–2055	−8.399	0.275	−4.404	0.094	−0.809	5.171	−3.595	−5.078
2050–2055/2055–2060	−8.686	0.098	−4.072	−0.037	−0.826	5.501	−3.246	−5.538
2055–2060/2060–2065	−8.434	0.011	−3.044	−0.664	−0.880	5.616	−2.165	−6.280
2060–2065/2065–2070	−7.893	0.009	−2.281	−0.855	−0.889	5.655	−1.392	−6.510
2065–2070/2070–2075	−6.356	0.071	−2.062	−0.307	−0.935	4.992	−1.127	−5.300
2070–2075/2075–2080	−4.922	0.164	−1.872	0.171	−0.954	4.340	−0.917	−4.169
2075–2080/2080–2085	−3.598	0.262	−1.521	0.253	−0.967	3.560	−0.554	−3.306
2080–2085/2085–2090	−2.504	0.349	−1.344	0.105	−0.987	2.601	−0.357	−2.496
2085–2090/2090–2095	−1.815	0.421	−1.065	−0.056	−0.951	2.065	−0.114	−2.121
2090–2095/2095–2100	−0.712	0.470	−0.758	−0.120	−0.974	1.278	0.216	−1.398
2015–2020/2095–2100	*−104.462*	*14.563*	*−57.203*	*−13.583*	*−13.794*	*62.032*	*−43.409*	*−75.615*

Source Author's calculations

death rate, *ad*. Table 3.5 suggests that the change in the age-independent birth rate; in the age structure effects on the birth rate; and in the age structure effects on the death rate are projected to contribute to decreasing the net addition to the population, whereas the changes in the age-independent death rate and in the population size are projected to contribute to increasing the net addition to the population in different quinquennials.

Table 3.5 also suggests that the net addition to the population of the country is projected to decrease by more than 104 million during the period 2015 through 2100. During 2015–2020, the net addition to the population is projected to be around 77 million, whereas the net addition to the population during 2095–2100 is projected to be around −28 million, which means a decrease, not increase, in the net addition to the population of the country during the quinquennials. The projected change in the net addition to the population during 2095–2100 as compared to the change during 2015–2020 is attributed to projected decrease of 57.2 million (54.8%) as a result of decrease in fertility; projected decrease of 62 million (59.4%) as a result of change in the age structure effects on the death rate; projected decrease of 13.6 million (13%) as a result of change in the age structure effects on the birth rate; projected increase of 14.6 million (13.9%) as a result of increase in population size; and projected increase of 13.8 million (13.2%) as a result of decrease in mortality. The projected decrease in the net addition to the population as a result of a decrease in fertility, the change in the age structure effects on the birth rate and the change in the age structure effects on the death rate will be around 133 million. On the other hand, the projected increase in the net addition to the population as a result of an increase in population size and a decrease in mortality will be more than 28 million so that the net decrease in the net addition to the population during 2095–2100 as compared to that during 2015–2100 will be more than 104 million.

Alternatively, the intrinsic population growth rate or the growth rate attributed to the difference between age-independent birth rate and age-independent death rate is projected to decrease from 2.479 per 1000 population during 2015–2020 to −2.362 per 1000 population during 2095–2100. The intrinsic population growth rate will turn negative during 2040–2045 when the age-independent death rate will become higher than the age-independent birth rate. However, the natural population growth rate will turn negative only during 2075–2080. The decrease in the intrinsic population growth rate is expected to result in a decrease of more than 43 million in the net addition to the population between 2015 and 2100. On the other hand, the decrease in the growth rate attributed to the change in population age structure is expected to result in a decrease of around 76 million in the net addition to the population so that the decrease in the natural population growth rate will result in a decrease of more than 119 million in the net addition to the population during 2095–2100 as compared to that during 2015–2020. This implies that the net addition to the population during 2095–2100 should be less than that during 2015–2020 by more than 32 million. However, the population in 2095 is projected to be higher than the population in 2015 by around 15 million so that the actual decrease in the net addition to the population during the quinquennials 2095–2100 will be less by around 104 million than the net addition to the population during 2015–2020 (Table 3.6).

Table 3.6 Decomposition of the increase in population of India: 2015–2100

Period	Increase in population (million)	Increase attributed to population size (million)	Increase attributed to age-independent birth rate (million)	Increase attributed to age structure effects on birth rate (million)	Increase attributed to age-independent death rate (million)	Increase attributed to age structure effects on death rate (million)	Increase attributed to the intrinsic component of growth (million)	Increase attribute to age component of growth (million)
2015–2020	76.775	76.775	76.775	76.775	76.775	76.775	0	0
2020–2025	71.109	80.738	70.921	74.103	75.936	78.716	−5.016	−4.613
2025–2030	63.832	83.878	64.638	71.971	75.165	81.490	−10.526	−9.519
2030–2035	54.394	86.188	58.108	69.333	74.409	84.826	−16.301	−15.494
2035–2040	43.700	87.724	52.016	66.549	73.682	88.907	−21.666	−22.358
2040–2045	34.170	88.665	46.636	65.012	72.924	93.219	−26.288	−28.206
2045–2050	25.631	89.208	41.995	64.609	72.153	98.028	−30.158	−33.419
2050–2055	17.232	89.483	37.591	64.703	71.344	103.200	−33.753	−38.497
2055–2060	8.547	89.581	33.519	64.666	70.518	108.701	−36.999	−44.035
2060–2065	0.113	89.592	30.475	64.001	69.639	114.317	−39.164	−50.315
2065–2070	−7.780	89.601	28.194	63.146	68.750	119.971	−40.556	−56.825
2070–2075	−14.136	89.672	26.132	62.839	67.815	124.964	−41.683	−62.125
2075–2080	−19.058	89.836	24.260	63.010	66.861	129.303	−42.600	−66.294
2080–2085	−22.657	90.098	22.739	63.263	65.893	132.863	−43.154	−69.600
2085–2090	−25.161	90.446	21.395	63.368	64.906	135.464	−43.511	−72.096
2090–2095	−26.975	90.867	20.330	63.312	63.956	137.529	−43.625	−74.217
2095–2100	−27.687	91.338	19.572	63.192	62.982	138.807	−43.409	−75.615
2015–2100	−104.462	14.563	−57.203	−13.583	−13.794	62.032		

Source Author's calculations

Finally, it is projected that India's population will increase by around 390 million or by almost 30% between 2015 and 2100. This increase will be the result of a decrease of more than 518 million as a result of decrease in the intrinsic population growth rate; a decrease of more than 723 million as a result of change in population age structure; and an increase of more than 1494 million as a result of increase in population size. This means that the decrease in fertility and mortality and the change in the population age structure will contribute towards decrease in the size of the population stock of the country in the coming years. However, these population decreasing effects will be offset by the population increasing effect attributed to the increase in population size so that the population of the country is projected to increase by almost 396 million between 2015 and 2065. After 2065, the population of the country will start decreasing so that the net addition to the population of the country during 2015–2100 will be around 390 million.

3.5 Conclusions

This chapter has attempted to quantify the contribution of the change in fertility, mortality and population age and sex structure to the projected increase in India's population stock during the period 2015–2100. The analysis suggests that the projected increase in India's population will be driven by its very large population size as the transition in fertility and mortality and the change in population age structure will contribute to decrease, instead increase, the population of the country in the coming years. The analysis also implies that the effects of the decrease in the intrinsic population growth rate and the decrease in the natural population growth rate attributed to the change in population age and sex structure on future population growth will be off-set by the momentum of growth resulting from the size of the population of the country. The effect of this growth momentum can be offset only by hastening the pace of fertility transition. However, the country has not been able to achieve fertility targets set in the past. The National Population Policy 2000 had aimed at achieving the replacement fertility by 2010 so as to stabilise population growth by 2040 (Government of India 2000). This target could not be achieved as the total fertility rate in the country was above the replacement level as late as 2015 as revealed through the sample registration system and the National Family Health Survey 2015–2016 (Government of India 2016; Government of India, no date). The National Health Policy 2017 has now set the target of achieving the replacement fertility at national and sub-national level by the year 2025 (Government of India 2017). The United Nations 2030 Sustainable Development Agenda has also set a target of achieving replacement fertility by 2030 (United Nations 2015). However, the medium or the most likely variant of the population projections prepared by the United Nations Population Division suggest that there is little possibility of achieving this target unless there is a comprehensive reinvigoration of fertility reduction efforts in the country. It is obvious that slower than expected fertility transition in the country in the past will have implications for the future population growth. However, there has been a sense of complacency in fertility reduction and population stabilisation efforts in the country in the recent past as is reflected through the decrease in the contraceptive prevalence rate from 56.3% in 2005–2006 to 53.5% during 2015–2016 (Government of India, no date). It was more than 17 years ago that a National Population Policy was announced in 2000. However, despite the fact that the goals of the National Population Policy 2000 could not be realised, there is little initiative to formulate a new population policy taking into account the current and future demographic prospects.

References

Andreev K, Kantorova V, Bongaarts J (2013) Demographic components of future population growth. New York, United Nations, Population Division. Technical Paper No. 2013/3

Bongaarts J (1994) Population policy options in the developing world. Science 263:2–7

Bongaarts J (2009) Human population growth and the demographic transition. Philos Trans R Soc B Biol Sci 364(1532):2985–2990

Bongaarts J, Bulatao RA (1999) Completing the demographic transition. Popul Dev Rev 25(3):515–529

Chaurasia AR, Gulati SC (2008) India: the state of population 2007. National Commission on Population and Oxford University Press, New Delhi

Frejka T (1973) The future of population growth: alternative paths to equilibrium. Wiley, New York

Frejka T (1981) Long-term prospects for world population growth. Popul Dev Rev 7(3):489–511

Government of India (no date) National Family Health Survey-4 2015-16. India Fact Sheet. Mumbai, International Institute for Population Sciences

Government of India (2000) National population policy 2000. Ministry of Health and Family Welfare, New Delhi

Government of India (2006) Population projections for India and States 2001-2016. Report of the Technical Group on Population Projections Constituted by the National Commission on Population. Registrar General and Census Commissioner, India, New Delhi

Government of India (2014) Census of India 2011. Report on post enumeration survey. Registrar General and Census Commissioner, India, New Delhi

Government of India (2016) Sample registration system statistical report 2015. Ministry of Home Affairs, Office of the Registrar General and Census Commissioner, New Delhi

Government of India (2017) National health policy 2017. Ministry of Health and Family Welfare, New Delhi

Horiuchi S (1995) The cohort approach to population growth: a retrospective decomposition of growth rates for Sweden. Popul Stud 49(1995):147–163

Keyfitz N (1971) On the momentum of population growth. Demography 8(1):71–80

Preston SH, Heuveline P, Guillot M (2001) Demography. Measuring and modeling population processes. Blackwell Publishers, Oxford

Schoen R, Kim YJ (1991) Convergence towards stability is a fundamental principle of population dynamics. Demography 28(3):455–466

United Nations (2015) Transforming our world: the 2030 agenda for sustainable development. United Nations, New York

United Nations (2017a) World population prospects. The 2017, key findings and advance tables. Department of Economic and Social Affairs. Population Division, New York

United Nations (2017b) World population prospects. The 2015 revision, methodology of the united nations population estimates and projections. Department of Economic and Social Affairs. Population Division, New York

Chapter 4
Fertility Transition, 1990–2010

Abstract This chapter analyses fertility transition in India during 1990–2010 on the basis of a fertility transition index that takes into account transition in both birth limitation and birth planning in the context of National Population Policy. The present analysis highlights the limitations of fertility transition efforts in India. Fertility transition efforts in the country remain confined to birth limitation at the cost of birth planning. There was evidence of transition in birth planning in the country during the 1990s but this transition either stagnated or reversed during the 2000s. The current pattern of fertility transition in India contributes little towards minimising the effect of momentum on future population growth.

4.1 Introduction

Fertility transition in India continues to be off the track. The National Population Policy 2000 had aimed at achieving the replacement level fertility (TFR = 2.1) by the year 2010 (Government of India 2000). This goal could not be achieved. The National Rural Health Mission, launched in 2005, aimed at achieving the replacement level fertility by the year 2012 (Government of India 2005). This goal could also not be achieved. The XII Five-year Development Plan (2012–2017) aimed at achieving the replacement fertility by the year 2017 (Government of India 2013a) but the National Health Policy 2017 now aims at achieving the replacement fertility by 2025 (Government of India 2017). According to the Sample Registration System, TFR in India was 2.2 per women of reproductive age during the period 2015-2017 (Government of India no date) while the United Nations Population Division has estimated a TFR of 2.44 during the period 2010–2015 (United Nations 2017). The United Nations projects that India is least likely to achieve replacement fertility before 2030. TFR in India decreased from around 2.9 in 2004–2006 to 2.2 in 2015-2017, but the total marital fertility rate remained unchanged during this period (Government of India no date).

An implication of slower than expected fertility transition in India is that the country will not be able to achieve the stable population by 2045 as stipulated in the National Population Policy 2000. Successful implementation of the National

© Springer Nature Singapore Pte Ltd. 2020
A. R. Chaurasia, *Population and Sustainable Development in India*,
https://doi.org/10.1007/978-981-32-9212-3_4

Population Policy 2000 would have resulted in restricting India's population growth to 1107 million by 2010. Against this expectation, the population enumerated at the 2011 population census was 1211 million and the post-enumeration survey suggested a net omission rate of 23 persons per 1000 population (Government of India 2014a). India's population was projected to increase to 1178.9 million by the year 2011 on the basis of the results of the 1991 population census (Government of India 1997), and to 1192.5 million on the basis of the results of the 2001 population census (Government of India 2007). This means that the actual population growth in India during 1991–2011 has been almost 66 million more than what was projected on the basis of the results of the 1991 population census.

The National Population Policy 2000 calls for promoting, vigorously, the 'small family norm' and delayed marriages for girls—not earlier than 18 years and, preferably, after 20 years of age—to reduce fertility and stabilise population growth at a level consistent with the requirements of sustainable economic growth, social development and environmental protection. The Policy contemplates an increase in the proportion of first- and second-order live births to women at least 20 years of age or, equivalently, a decrease in the proportion of births to women younger than 20 years of age and a decrease in the proportion of third and higher-order births. The promotion of two-child family and delayed childbearing constitute the two different dimensions of fertility transition. Fertility decreases when couples limit the number of children they want. Fertility also decreases when the entry into marital union is delayed and there is an increase in the spacing between marriage and the first birth, and between successive births. Birth limitation has direct impact on the completed fertility but birth planning does not, although it has implications for population growth. It is well known that even when the replacement fertility (TFR = 2.1) is achieved, population continues to grow because of the in-built momentum in the population resulting from young age structure (Keyfitz 1971; Frejka 1982; Merrick 1986). The effect of population momentum cannot be eliminated. It can, at best, be minimised either by lowering further the completed fertility or by increasing the mean age at childbearing. Theoretically, completed fertility can be decreased to a level at which the birth rate is equal to the death rate. This, however, implies that many couples would have only one child (Bongaarts 1994) which is not a feasible proposition in India, given the persistence of the above average infant and child mortality in the country. The second, more feasible, option is to increase the mean age at childbearing by increasing the age at marriage, especially of females, delaying childbearing by increasing the interval between marriage and first birth and properly spacing successive births. Population momentum matters for India as future population growth in the country will largely be the result of population momentum and not because of high fertility (Chaurasia and Gulati 2008; Chaurasia 2016).

Birth planning has implications for the health and survival of mothers and children also (CATALYST Consortium 2009; Conde-Agudelo and Belizan 2000; Conde-Agudelo et al. 2012; Davanzo et al. 2004; Cleland and Sathar 1984; Hobcraft et al. 1985). The critical role of birth planning as one of the most cost-effective, high-yield interventions to improve the survival and health of mothers and children is well known (UNDP 2003; WHO 2005; Moreland and Talbird 2006; Smith et al. 2009).

The length of the preceding birth interval is a major determinant of infant and early childhood mortality (Fotse et al. 2013). Birth planning also contributes to lowering the demand for 'bonus' children and thus contributes to promoting birth limitation. High infant and child mortality also keeps the replacement level fertility higher than the universally accepted norm of TFR = 2.1 (Espenshade et al. 2003). Birth planning, therefore, has an additive impact on birth limitation.

It is imperative from the forgoing discussions that any analysis of fertility transition should consider transition in both birth planning and birth limitation. Fertility transition is most commonly measured and analysed in terms of TFR, which is an indicator of completed fertility. Any analysis of fertility transition based on TFR, therefore, is biased towards transition in birth limitation. Another limitation of TFR as a measure of fertility transition is that it is associated with tempo effects or the effects of the timing of births (Government of India 2013b). A couple may produce two births within two years of marriage and stops childbearing. Another couple produces first birth after three years of marriage and manages a spacing of three years between first and second birth and then stops childbearing. In both cases, the TFR will be 2.1 but the implications of the reproductive behaviour of the first couple to population growth and health of women and children are radically different from the reproductive behaviour of the second couple. The National Population Policy 2000 discourages the reproductive behaviour adopted by the first couple and encourages adoption of the reproductive behaviour of the second couple in the quest towards universalisation of the 'small family' norm.

In this chapter, we analyse fertility transition in India during 1991–2011 on the basis of a fertility transition index that takes into account the transition in both the dimensions of fertility transition—birth planning and birth limitation. The period 1991–2011 has been a period of profound changes in the basic philosophy of the official family planning efforts in India. In 1992, the National Family Welfare Programme, the main plank of fertility reduction efforts in the country, was comprehensively revamped. In 1996, the target-based approach of the implementation of the official family planning programme was discontinued to give way for a 'target free' or community needs-based approach of programme implementation. A new National Population Policy was announced in the year 2000 and the National Commission on Population was constituted in the same year. The National Rural Health Mission was launched in 2005 and the National Family Welfare (Planning) Programme was subsumed into the Reproductive and Child Health component of the Mission so that the separate Department of Family Welfare within the Ministry of Health and Family Welfare of the Government of India that spearheaded fertility transition efforts in the country for more than 50 years sieged to exist (Chaurasia and Singh 2014). Family planning in India is now seen as a strategy to reduce maternal and child mortality rather than as an approach to reduce fertility and control population growth. There is, however, little empirical evidence to suggest how these changes have influenced fertility transition in the country.

The analysis is based on the fertility transition index that takes into account transition in both, birth planning and birth limitation (Chaurasia 2013a). The index follows the numerator-based approach of measuring and monitoring fertility transition (Ravenholt and Frederiksen 1968; Reynolds 1972; Chidambaram 1965; Balasubramanian 1972; Bertrand et al. 1994). The numerator-based approach has been argued to be more sensitive to short-term changes in individual fertility behaviour than completed fertility measures, such as total fertility rate and birth rate (Ryder 1982; Srinivasan and Freymann 1990). The numerator-based approach is based on the concept of 'undesired' fertility. The National Population Policy 2000 emphasises that live births to women younger than 20 years and third and higher-order live births should be avoided in the efforts towards the universal realisation of the 'small family' as the strategy to stabilise population. From the policy perspective, this means that births to women below 20 years of age and third and higher-order births should be regarded as 'undesired' births. This also means that fertility transition should be measured and analysed in terms of the reduction in the proportion of 'undesired' births.

More specifically, the objectives of this chapter are as follows. We first outline the fertility transition index that takes into account the proportion of births to women below 20 years of age and the proportion of third and higher-order births. Next, we estimate the fertility transition index for the country and for its constituent states/Union Territories and districts for the year 1990, 2000 and 2010 on the basis of the data available through 1991, 2001 and 2011 population census. The fertility transition index for states/Union Territories and districts have then been used to analyse the convergence in fertility transition across the states/Union Territories. The paper also analyses the relative contribution of the change in the proportion of births to women below 20 years of age and in the proportion of third and higher-order births to the change in the fertility transition index. Finally, the paper analyses defining characteristics of districts at different stages of fertility transition.

4.2 Fertility Transition Index

The fertility transition index (f) used in the present analysis is based on the distribution of live births by the age of the mother and the order of the birth (Chaurasia 2013a, b). Let B_{ij} denotes the number births of order j to women aged i in a year. Then the total number of births in a year, B, is equal to

$$B = \sum_i \sum_j B_{ij} \tag{3.1}$$

$$B = \sum_{i<20} \sum_{j<3} B_{ij} + \sum_{i<20} \sum_{j\geq 3} B_{ij} + \sum_{i\geq 20} \sum_{j<3} B_{ij} + \sum_{i\geq 20} \sum_{j\geq 3} B_{ij} \tag{3.2}$$

$$B = \sum_{i\geq 20} \sum_{j<3} B_{ij} + \sum_{i<20} \sum_{j} B_{ij} + \sum_{i} \sum_{j\geq 3} B_{ij} - \sum_{i<20} \sum_{j\geq 3} B_{ij} \tag{3.3}$$

Dividing both the sides by total number of births B, we get

$$1 = b_{i \geq 20, j<3} + b_{i<20} + b_{j<3} - \mathrm{b}_{i<20, j \geq 3}, \text{ or}$$
$$b_{i \geq 20, j<3} = 1 - \left(b_{i<20} + b_{j<3} - b_{i<20, j \geq 3}\right) \tag{3.4}$$

where

$b_{i \geq 20, j<3}$ = proportion of births to women aged 20–49 years and birth order <3.
$b_{i<20}$ = proportion of births to women below 20 years of age.
$b_{j<3}$ = proportion of third and higher-order births irrespective of the age of women.
$b_{i<20, j \geq 3}$ = proportion of third and higher-order births to women below 20 years of age.

Assuming that $b_{i<20, j<3}$ is small in comparison to $b_{i<20}$ and $b_{j<3}$, then the fertility transition index, *f*, may be defined as

$$f = 1 - \left(b_{i<20} + b_{j \geq 3}\right) \tag{3.5}$$

When $f = 1$, all the reported births in a year are first and second-order births to women aged at least 20 years. On the other hand, f will be negative when all births in a year are third and higher-order births to women aged less than 20 years. However, as f increases, $b_{i<20, j \geq 3}$ becomes less and less important. When $f = 1$, both $b_{i<20}$ and $b_{j<3}$ and hence $b_{i<20, j \geq 3} = 0$. For all practical purposes, therefore, it can be assumed that f ranges between 0 and 1. The index f captures transition in both birth planning and birth limitation. The progression from second to third birth is a crucial component of fertility change (United Nations 1997) and the decrease in completed fertility is reflected in the decrease in the proportion of third and higher-order births (Prasartakul et al. 1987; Srinivasan et al. 1992; Singh 2002). On the other hand, the latter is the age at first birth, the lower is the fertility. In addition, a decreasing proportion of births to women aged younger than 20 years is a reflection of the rising age at first birth (Sivakumar 2000). An increase in the age at first birth and increased spacing between successive births is therefore likely to result in a decrease in the proportion of births to women younger than 20 years of age.

Based on the fertility transition index *f*, fertility in a population can be classified as: at very early stage of transition if $f < 0.50$; at early stage of transition if $0.50 \leq f < 0.60$; at the middle stage of transition if $0.60 \leq f < 0.70$; at advanced stage of transition if $0.70 \leq f < 0.80$; and at very advanced stage of transition if $f \geq 0.80$. A fertility transition index $f < 0.50$ implies that more than half of the births in a year are essentially 'undesired' in the sense that these births are either births to women aged less than 20 years or are third and higher-order births. On the other hand, a fertility transition index $f \geq 0.80$ implies that 20% and less births are 'undesired'. It is also obvious that the rapid the rate of the increase in *f*, the faster the pace of transition in fertility.

It is obvious that the change in the fertility transition index f over time is the result of the change in the proportion of births to women below 20 years of age ($b_{i<20}$) and in the proportion of third and higher-order births ($b_{j \geq 3}$). In the present case, the change in the fertility transition index, *f*, between 1990 and 2010 can be decomposed as:

$$f_{2010} - f_{1990} = (f_{2010} - f_{2000}) + (f_{2000} - f_{1990})$$

Now

$$f_{2010} - f_{2000} = \left(b_{i<20,\,2000} - b_{i<20,\,2010}\right) + \left(b_{j>3,\,2000} - b_{j>3,\,2010}\right)$$

and

$$f_{2000} - f_{1990} = \left(b_{i<20,\,1990} - b_{i<20,\,2000}\right) + \left(b_{j>3,\,1990} - b_{j>3,\,2000}\right)$$

Substituting, we get

$$f_{2010} - f_{1990} = \left(b_{i<20,\,2000} - b_{i<20,\,2010}\right) + \left(b_{j>3,\,2000} - b_{j>3,\,2010}\right) + \left(b_{i<20,\,1990} - b_{i<20,\,2000}\right) + \left(b_{j>3,\,1990} - b_{j>3,\,2000}\right) \quad (3.6)$$

In other words, the change in the fertility transition index, *f*, between 1990 and 2010 is the algebraic sum of the change in the proportion of births to women below 20 years during 1990–2000 and during 2000–2010 and the change in the proportion of third and higher-order births during 1999–2000 and during 2000–2010. This decomposition makes it possible to analyse the basic features of fertility transition during 1999–2000 as compared to those during 2000–2010 or before and after the announcement of the National Population Policy 2000. In addition, it is also possible to analyse the relative contribution of the two dimensions of fertility transition—the dimension of birth planning and the dimension of birth limitation—to the transition in fertility as reflected through the change in the fertility transition index *f*.

4.3 Data

The analysis is based on the number of live births reported during one year prior to India's 1991, 2001 and 2011 population census and, therefore, refers to years 1990, 2000 and 2010, respectively. These live births are classified by the age of the mother at the time of the birth and the order of the birth and are available for the country and for its constituent states/Union Territories and districts as they existed at the time of the respective population census. Data available through the population census in India are generally associated with errors of under-enumeration and age misreporting. The post-enumeration survey carried out after the 2011 population census has indicated that the inconsistency between census enumeration and post-enumeration survey is, in general, low (Government of India 2014a). Live births reported at the population census and their classification by age of the mother at the time of the birth and by the order of the birth appear to be fairly accurate when compared with the data from India's Sample Registration System. According to the Sample Registration System, fertility of women aged 15–19 years accounted for around 6.8% of the cumulative fertility, whereas the proportion of third and higher-order births accounted for around 27.6% of the total live births in the year 2011 (Government of India 2014b). The corresponding proportions based on 2011 census are 6.4 and 31.6%, respectively, which shows that the data available from the population census are largely accurate. The data available through the population census are also associated with the errors of missing birth order and missing age of the mother at the time of birth. These errors were adjusted by applying the methods protocol of the human fertility database (Jasilioniene et al. 2012).

4.4 Fertility Transition in India

The index of fertility transition f in India is estimated to be 0.619 for the year 2010, which implies that the country was in the middle of fertility transition, circa 2010 (Table 4.1). About 32% of the births reported during the year preceding the 2011 population census were third and higher-order births, whereas births to women below 20 years of age were around 6% so that around 38% of the births were 'undesired'. The differentials in fertility transition across residence, social class and religion are quite strong in the country. Fertility transition was more advanced in urban than in rural population according to the data available through 2011 population census. Similarly, fertility transition varied widely across religious communities and social classes. The Jain community was the only religious community where fertility was at very advanced stage of transition whereas in the Muslim community, fertility was at an early stage of transition. In Hindu, Christian and Buddha communities, fertility appeared to be in the middle of transition, whereas in the Sikh community, it was at the advanced stage. Fertility was also at the early stage of transition in both Scheduled Tribes and Scheduled Castes but in the middle stage in other social classes. Table 4.1 also suggests that fertility transition is strongly related to the educational status of the mother. Among the illiterate mothers, fertility was at very early stage of transition whereas in mothers with at least graduate level education, fertility transition appeared to be almost complete. Less than 10% of births to these women during 2010 were 'undesired'.

Fertility transition varied widely across the states and Union Territories of the country. Fertility was at very advanced stage of transition in Goa, Kerala and Puducherry, circa 2010. In these states, less than 20% of births were 'undesired' (Table 4.2). In contrast, in Meghalaya and Nagaland, more than half of the births in 2010 were 'undesired' indicating that fertility, in these states, was at very early stage of transition. There are eight states/Union Territories, where fertility was at advanced stage of transition in 2010 but in nine states/Union Territories, it was at early stage of transition. This leaves 13 states/Union Territories where fertility was at the middle stage of transition. By comparison, fertility in 1990 was at advanced stage of transition in three states/Union Territories; in the middle of transition in five states/Union Territories; and at very early stage of transition in 26 states/Union Territories. There was no state/Union Territory where fertility was at very advanced

Table 4.1 Fertility transition index (f), proportion of births to women <20 years of age and proportion of third and higher-order births in India by residence, social class, religion and mother's education, 1990–2010

Population		Fertility transition index (f)			Proportion of births to women <20 years of age			Proportion of 3rd and higher order births		
		1990	2000	2010	1990	2000	2010	1990	2000	2010
Total population		0.418	0.469	0.619	0.085	0.066	0.064	0.497	0.466	0.316
Residence	Rural	0.401	0.439	0.587	0.088	0.070	0.070	0.511	0.491	0.343
	Urban	0.483	0.575	0.708	0.073	0.051	0.049	0.444	0.374	0.243
Social class	Scheduled Castes	0.398	0.414	0.587	0.096	0.075	0.071	0.506	0.510	0.342
	Scheduled Tribes	0.413	0.394	0.527	0.079	0.066	0.077	0.508	0.541	0.396
	Others	0.422	0.493	0.640	0.083	0.063	0.061	0.494	0.444	0.299
Religion	Hindu	0.428	0.482	0.634	0.086	0.067	0.064	0.486	0.450	0.302
	Muslim	0.318	0.363	0.529	0.090	0.067	0.070	0.591	0.570	0.401
	Christian	0.540	0.580	0.675	0.038	0.031	0.042	0.422	0.390	0.284
	Sikh	0.566	0.607	0.775	0.039	0.032	0.031	0.395	0.361	0.195
	Buddhist	0.407	0.525	0.684	0.083	0.053	0.064	0.510	0.422	0.252
	Jain	0.644	0.736	0.812	0.034	0.015	0.018	0.322	0.248	0.170
	Others	0.338	0.398	0.551	0.073	0.058	0.072	0.590	0.545	0.377
Mother's education	Illiterate	0.354	0.350	0.472	0.085	0.061	0.051	0.561	0.589	0.478
	Literate below primary	na	0.461	0.558	na	0.086	0.081	na	0.453	0.361
	Primary below middle	0.455	0.531	0.628	0.100	0.094	0.088	0.445	0.375	0.284
	Middle below matriculate	0.568	0.623	0.690	0.100	0.086	0.094	0.331	0.290	0.217
	Matriculate below graduate	0.706	0.741	0.781	0.055	0.050	0.074	0.238	0.209	0.145
	Graduate and above	0.845	0.881	0.920	0.006	0	0	0.149	0.112	0.080

Source Author's calculations

Table 4.2 Fertility transition index, proportion of births to women below 20 years of age and proportion of third and higher order births in states and Union Territories

State/Union territory	Fertility transition index (*f*)			Proportion of births to women <20 years of age			Proportion of 3rd and higher order births		
	1990	2000	2010	1990	2000	2010	1990	2000	2010
Andaman and Nicobar Islands	0.501	0.648	0.758	0.069	0.057	0.069	0.430	0.295	0.173
Andhra Pradesh	0.424	0.572	0.709	0.152	0.144	0.090	0.424	0.284	0.201
Arunachal Pradesh	0.276	0.374	0.536	0.089	0.054	0.060	0.635	0.572	0.404
Assam	0.357	0.426	0.582	0.075	0.058	0.079	0.569	0.516	0.339
Bihar	0.342	0.360	0.542	0.081	0.060	0.060	0.577	0.579	0.398
Chandigarh	0.552	0.596	0.751	0.039	0.035	0.027	0.409	0.368	0.222
Chhattisgarh	0.401	0.433	0.626	0.091	0.054	0.055	0.508	0.513	0.319
Dadra and Nagar Haveli	0.427	0.442	0.649	0.069	0.068	0.072	0.504	0.490	0.279
Daman and Diu	0.499	0.617	0.753	0.052	0.037	0.048	0.449	0.345	0.198
Delhi	0.485	0.557	0.726	0.051	0.038	0.029	0.464	0.405	0.245
Goa	0.623	0.729	0.833	0.022	0.022	0.022	0.355	0.249	0.144
Gujarat	0.499	0.539	0.663	0.043	0.041	0.055	0.458	0.420	0.282
Haryana	0.456	0.490	0.670	0.085	0.076	0.056	0.459	0.435	0.274
Himachal Pradesh	0.512	0.616	0.765	0.058	0.029	0.032	0.429	0.354	0.203
Jammu and Kashmir	na	0.407	0.540	na	0.027	0.026	na	0.567	0.434
Jharkhand	0.345	0.384	0.552	0.100	0.076	0.081	0.555	0.540	0.367
Karnataka	0.453	0.560	0.665	0.095	0.081	0.063	0.451	0.360	0.272
Kerala	0.666	0.733	0.806	0.061	0.049	0.040	0.273	0.218	0.154
Lakshadweep	0.356	0.473	0.666	0.063	0.037	0.020	0.581	0.490	0.313

(continued)

Table 4.2 (continued)

State/Union territory	Fertility transition index (*f*)			Proportion of births to women <20 years of age			Proportion of 3rd and higher order births		
	1990	2000	2010	1990	2000	2010	1990	2000	2010
Madhya Pradesh	0.360	0.404	0.531	0.097	0.067	0.065	0.543	0.529	0.404
Maharashtra	0.474	0.572	0.674	0.078	0.056	0.067	0.449	0.371	0.259
Manipur	0.351	0.482	0.674	0.037	0.027	0.045	0.612	0.491	0.281
Meghalaya	0.292	0.336	0.470	0.050	0.039	0.067	0.658	0.625	0.463
Mizoram	0.348	0.452	0.548	0.033	0.045	0.062	0.619	0.503	0.390
Nagaland	0.286	0.322	0.498	0.033	0.026	0.042	0.681	0.652	0.460
Orissa	0.426	0.513	0.694	0.064	0.045	0.056	0.509	0.443	0.251
Puducherry	0.605	0.710	0.833	0.072	0.063	0.043	0.323	0.227	0.125
Punjab	0.556	0.595	0.774	0.038	0.029	0.029	0.406	0.376	0.196
Rajasthan	0.376	0.397	0.514	0.090	0.079	0.072	0.534	0.524	0.414
Sikkim	0.373	0.477	0.693	0.056	0.077	0.102	0.571	0.445	0.205
Tamil Nadu	0.563	0.648	0.765	0.067	0.058	0.051	0.370	0.294	0.184
Tripura	0.396	0.537	0.674	0.089	0.094	0.128	0.515	0.369	0.199
Uttar Pradesh	0.331	0.349	0.560	0.069	0.043	0.049	0.599	0.608	0.391
Uttarakhand	0.419	0.469	0.676	0.052	0.036	0.038	0.528	0.495	0.286
West Bengal	0.416	0.499	0.630	0.120	0.113	0.134	0.464	0.388	0.235

Source Author's calculations

stage of transition in 1990. Fertility transition could not be analysed in Jammu and Kashmir as no population census was carried out in the state in 1991.

Fertility transition also varied widely across districts. In 1990, fertility was at very early stage of transition in 345 of 452 districts (76.3%) as they existed at the time of 1991 population census. By comparison, there were only 75 out of 640 districts (11.7%) in 2010 where fertility was at very early stage of transition (Table 4.3). In 1990, there were only 9 (2%) districts where fertility was at advanced or very advanced stage of transition compared to 185 (28.9%) districts in 2010. The fertility transition index f was the lowest in district Badgam of Jammu and Kashmir in 2010 but the highest in district Pathanamthitta of Kerala. In 1990 also, district Pathanamthitta of Kerala had the highest fertility transition index in the country but the index was the lowest in district Upper Subansiri of Arunachal Pradesh. The inter-district distribution of the fertility transition index in 1990 was highly skewed to the left, meaning that a large proportion of districts were having fertility transition index lower than the average. However, in 2010, the inter-district distribution of the fertility transition index has turned very close to normal. Similarly, the inter-district distribution turned from a leptokurtic one in 1990 to a platykurtic one in 2010, which means that the dispersion in the fertility transition index across districts had increased over time.

Data available from the population census indicate that the country has moved from the very early stage of fertility transition in 1990 to the middle stage of fertility transition in 2010. The pace of fertility transition was, however, slower in the

Table 4.3 Inter-district variation in fertility transition index (f), 1990–2010

Fertility transition index	Year		
	1990	2000	2010
Number of Districts			
Less than 0.500	345	359	75
0.500–0.600	74	132	222
0.600–0.700	24	81	158
0.700–0.800	4	13	155
0.800 and above	5	8	30
N	452	593	640
Summary Measures			
Minimum	0.235	0.248	0.360
Q1	0.350	0.370	0.539
Median	0.400	0.458	0.616
Q3	0.491	0.564	0.715
Maximum	0.855	0.865	0.918
Skewness	1.184	0.624	0.151
Excess kurtosis	1.731	−0.198	−0.697

Source Author's calculations

rural population but faster in the urban population. As a result, fertility in the rural population in 2010 was in the early stage of transition but at the advanced stage of transition in the urban population. Among different religious communities, the pace of fertility transition was faster in Muslim as compared to the Hindu religious community, possibly because fertility transition in Hindus was relatively more advanced in 1990 as compared to Muslims. The pace of fertility transition was, however, the slowest in Christian and Jain religious communities as fertility in these communities was at fairly advanced stage of transition in 1990. The Jain community was the only community where fertility was at the middle stage of transition in 1990. Similarly, the pace of fertility transition was the slowest among mothers having at least graduate level of education simply because fertility transition in this sub-group of population was already at very advanced stage of transition in 1990.

Fertility transition was most rapid in Arunachal Pradesh during the period under reference but slowest in Kerala because the index f was the lowest in Arunachal Pradesh but the highest in Kerala in 1990. In Arunachal Pradesh, more than 46% births in 2010 were 'undesired' compared to less than 20% in Kerala. The inter-state variation in the index f was, however, the lowest in 1990 with a standard deviation of 0.098 compared to 0.112 in 2000 and 0.098 in 2010. The increase in the standard deviation of inter-state distribution of the index f suggests the absence of σ-convergence in fertility transition across the states/Union Territories during 1990–2010. σ-convergence occurs when the inter-state dispersion in the index f decreases over time. It appears that fertility transition across the states/Union Territories diverged during 1990–2000 but converged during 2000–2010.

On the other hand, the decrease in the fertility transition index f during 1990–2010 across the states/Union Territories has been found to be negatively correlated to the level of the index in 1990 which indicates the presence of β-convergence in fertility transition across the states/Union Territories. β-convergence implies that fertility transition during 1990–2010 was relatively faster in those states/Union Territories which were relatively at earlier stage of transition in 1990 as compared to states/Union Territories which were relatively at later stage of transition in 1990. Moreover, unlike σ-convergence, β-convergence in fertility transition across the states/Union Territories occurred during both 1990–2000 and 2000–2010, although the speed of convergence was more rapid during 2000–2010 than during 1990–2000. β-convergence is a necessary but not the sufficient condition for σ-convergence in transition in fertility across the states/Union Territories (Young et al. 2008).

The analysis of convergence in fertility across districts could not be carried out because the number of districts in the country increased from 452 at the time of 1991 population census to 593 at the 2001 population census and 640 at the 2011 population census. However, the standard deviation of the inter-district distribution of the index f was 0.105 in 1990 which increased to 0.121 in 2000 but decreased to 0.110 in 2010. Although the numbers of districts in 1990, 2000 and 2010 were different, yet it appears that inter-district dispersion in fertility transition increased during the period 1990–2010 mainly because of the increase in dispersion during 1990–2000 as inter-district dispersion decreased during 2000–2010.

Table 4.4 decomposes the change in the index *f* during 1990–2010 into the change in its four components. Almost three-fourths of the increase in the index *f* during 1990–2010 was due to the decrease in the proportion of third and higher-order births during 2000–2010. On the other hand, the decrease in the proportion of births to women below 20 years of age during 2000–2010 contributed to just about 1% of the increase in the index *f* during 1990–2010. Table 4.4 suggests that fertility transition in the country during 1990–2010 was primarily due to the transition in the birth limitation dimension of fertility transition as reflected through the decrease in the proportion of third and higher-order births and that too during 2000–2010 only. On the other hand, the contribution of the transition in the birth planning dimension of fertility transition as reflected through the decrease in the proportion of births to women below 20 years of age was, at best, marginal. Moreover, it may also be seen from the table that more than three-fourths of the increase in the index *f* during 1990–2010 was confined to the period 2000–2010 only.

The relative contribution of the four components of the change in the index *f* is found to vary by population sub-groups. In the rural population, only 20% of the increase in the index *f* was confined to the period 1990–2000 compared to more than 40% in the urban population. However, 90% of the increase in the index *f* is attributed to the decrease in the proportion of third and higher-order births. Similarly, fertility transition in most of the religious communities has also been confined primarily to the period 2000–2010, with the exception of Buddha and Jain communities. The Jain community is the only community where the increase in *f* was more rapid during 1990–2000 than during 2000–2010. However, in all religious communities, at least 90% of the increase in the index *f* was due to the decrease in the proportion of third and higher-order births.

On the other hand, fertility transition appears to have reversed in the Scheduled Tribes population and in illiterate mothers during 1990–2000. The relative importance of different components to the increase in the index *f*, however, varied by the educational status of the mother. Among illiterate mothers, almost 30% of the increase in the index *f* was attributed to the decrease in the proportion of births to women below 20 years of age, whereas the proportion of births to women below 20 years of age increased, instead decreased, among mothers with more than matriculate but below graduate level education. On the other hand, fertility transition was more rapid during 1990–2000 than during 2000–2010 in mothers with at least graduate level education.

The relative contribution of the four components of the change in the index *f* has also been different in different states/Union Territories (Table 4.5). In Bihar, Dadra and Uttar Pradesh, more than 90% of the increase in the index *f* during 1990–2010 was due to the decrease in the proportion of third and higher-order births during 2000–2010. In contrast, decrease in the proportion of third and higher-order births during 2000–2010 accounted for less than 50% increase in the index *f* in Goa, Andaman and Nicobar Islands, Andhra Pradesh, Tripura and Mizoram. In Kerala, Karnataka, Andhra Pradesh, Chhattisgarh and Madhya Pradesh, at least 15% of the increase in the index *f* was due to the decrease in the proportion of births to women below 20 years of age but in 10 states/Union territories, the proportion of births

Table 4.4 Decomposition of the change in fertility transition index (f) in India, 1990–2010

Population		Change in f during 1990–2010	Proportionate (%) change attributed to the						Transition in	
			Change in the proportion of births to women <20 years of age		Change in the proportion of 3rd and higher order births		Change during the time period			
			1990–2000	2000–2010	1990–2000	2000–2010	1990–2000	2000–2010	Birth planning	Birth limitation
Total		0.202	9.4	1.0	15.3	74.3	24.8	75.2	10.4	89.6
Social class	Scheduled Castes	0.189	11.1	2.1	−2.1	88.9	9.0	91.0	13.2	86.8
	Scheduled Tribes	0.114	11.4	−9.6	−28.9	127.2	−17.5	117.5	1.8	98.2
	Others	0.217	9.2	0.9	23.0	66.8	32.3	67.7	10.1	89.9
Religion	Hindu	0.206	9.2	1.5	17.5	71.8	26.7	73.3	10.7	89.3
	Muslim	0.210	11.0	−1.4	10.0	80.5	21.0	79.0	9.5	90.5
	Christian	0.134	5.2	−8.2	23.9	79.1	29.1	70.9	−3.0	103.0
	Sikh	0.208	3.4	0.5	16.3	79.8	19.7	80.3	3.8	96.2
	Buddhist	0.277	10.8	−4.0	31.8	61.4	42.6	57.4	6.9	93.1
	Jain	0.168	11.3	−1.8	44.0	46.4	55.4	44.6	9.5	90.5
	Others	0.214	7.0	−6.5	21.0	78.5	28.0	72.0	0.5	99.5

(continued)

Table 4.4 (continued)

Population		Change in f during 1990–2010	Proportionate (%) change attributed to the						Transition in	
			Change in the proportion of births to women <20 years of age		Change in the proportion of 3rd and higher order births		Change during the time period			
			1990–2000	2000–2010	1990–2000	2000–2010	1990–2000	2000–2010	Birth planning	Birth limitation
Mother's education	Illiterate	0.117	20.5	8.5	−23.9	94.9	−3.4	103.4	29.1	70.9
	Literate below middle	0.153	5.7	2.9	27.6	63.8	33.3	66.7	8.6	91.4
	Middle below matric	0.120	11.7	−6.7	34.2	60.8	45.8	54.2	5.0	95.0
	Matric below graduate	0.074	6.8	−32.4	39.2	86.5	45.9	54.1	−25.7	125.7
	Graduate and above	0.075	8.0	0	49.3	42.7	57.3	42.7	8.0	92.0

Source Author's calculations

Table 4.5 Decomposition of the change in fertility transition index (f) in states/Union Territories, 1990–2010

State/Union Territory	Change in f 1990–2010	Proportionate change								
		All	Proportion of births to women <20 years		Proportion of 3rd and higher order births		Period		Dimension	
		1990–2010	1990–2000	2000–2010	1990–2000	2000–2010	1990–2000	2000–2010	Birth planning	Birth limitation
Andaman and Nicobar Islands	0.5	100.0	4.6	−4.6	52.6	47.5	57.1	42.9	0.0	100.0
Andhra Pradesh	0.7	100.0	2.5	19.1	49.3	29.1	51.8	48.2	21.6	78.4
Arunachal Pradesh	0.9	100.0	13.6	−2.2	24.2	64.5	37.8	62.2	11.4	88.6
Assam	0.6	100.0	7.5	−9.2	23.3	78.3	30.9	69.1	−1.6	101.6
Bihar	0.6	100.0	10.6	0.3	−1.4	90.5	9.2	90.8	10.9	89.1
Chandigarh	0.4	100.0	1.8	4.2	20.4	73.7	22.1	77.9	6.0	94.0
Chhattisgarh	0.6	100.0	16.4	−0.5	−2.3	86.3	14.2	85.8	16.0	84.0
Dadra and Nagar Haveli	0.5	100.0	0.7	−2.0	6.0	95.3	6.7	93.3	−1.2	101.2
Daman and Diu	0.5	100.0	5.8	−4.4	40.7	57.9	46.5	53.5	1.4	98.6
Delhi	0.5	100.0	5.6	3.6	24.4	66.4	30.0	70.0	9.2	90.8
Goa	0.3	100.0	−0.3	0.0	50.4	49.8	50.1	49.9	−0.3	100.3
Gujarat	0.3	100.0	1.3	−8.4	23.2	83.9	24.5	75.5	−7.1	107.1

(continued)

Table 4.5 (continued)

State/Union Territory	Change in f 1990–2010	Proportionate change								
		All	Proportion of births to women <20 years		Proportion of 3rd and higher order births		Period		Dimension	
		1990–2010	1990–2000	2000–2010	1990–2000	2000–2010	1990–2000	2000–2010	Birth planning	Birth limitation
Haryana	0.5	100.0	4.5	9.2	11.4	74.9	15.9	84.1	13.7	86.3
Himachal Pradesh	0.5	100.0	11.4	−1.0	29.7	59.9	41.1	58.9	10.4	89.6
Jammu and Kashmir*	0.1	100.0	na	0.5	na	99.5	na	100.0	0.5	99.5
Jharkhand	0.6	100.0	11.8	−2.7	7.4	83.5	19.2	80.8	9.2	90.8
Karnataka	0.5	100.0	7.0	8.2	43.2	41.5	50.3	49.7	15.3	84.7
Kerala	0.2	100.0	8.4	6.6	39.5	45.5	47.9	52.1	15.0	85.0
Lakshadweep	0.9	100.0	8.4	5.3	29.3	57.0	37.7	62.3	13.7	86.3
Madhya Pradesh	0.5	100.0	17.2	1.3	8.3	73.2	25.5	74.5	18.5	81.5
Maharashtra	0.4	100.0	10.7	−5.4	38.7	56.1	49.4	50.6	5.3	94.7
Manipur	0.9	100.0	3.0	−5.6	37.6	64.9	40.7	59.3	−2.6	102.6
Meghalaya	0.6	100.0	6.0	−15.5	18.8	90.7	24.9	75.1	−9.5	109.5

(continued)

Table 4.5 (continued)

State/Union Territory	Change in *f* 1990–2010	Proportionate change								
		All	Proportion of births to women <20 years		Proportion of 3rd and higher order births		Period		Dimension	
		1990–2010	1990–2000	2000–2010	1990–2000	2000–2010	1990–2000	2000–2010	Birth planning	Birth limitation
Mizoram	0.6	100.0	−5.9	−8.3	57.9	56.3	52.0	48.0	−14.2	114.2
Nagaland	0.7	100.0	3.3	−7.7	13.9	90.5	17.2	82.8	−4.3	104.3
Orissa	0.6	100.0	7.4	−4.1	25.0	71.8	32.3	67.7	3.3	96.7
Puducherry	0.4	100.0	3.9	9.0	42.0	45.0	45.9	54.1	12.9	87.1
Punjab	0.4	100.0	4.2	0.0	13.8	82.0	18.0	82.0	4.2	95.8
Rajasthan	0.4	100.0	8.3	5.1	7.5	79.2	15.8	84.2	13.4	86.6
Sikkim	0.9	100.0	−6.7	−7.8	39.2	75.2	32.5	67.5	−14.4	114.4
Tamil Nadu	0.4	100.0	4.3	3.6	37.8	54.3	42.1	57.9	7.9	92.1
Tripura	0.7	100.0	−2.0	−12.0	52.8	61.1	50.9	49.1	−14.0	114.0
Uttar Pradesh	0.7	100.0	11.4	−2.4	−3.8	94.8	7.6	92.4	9.0	91.0
Uttarakhand	0.6	100.0	6.4	−0.8	13.1	81.3	19.4	80.6	5.6	94.4
West Bengal	0.5	100.0	3.4	−9.9	35.4	71.1	38.8	61.2	−6.5	106.5

Source Author's calculations

to women below 20 years of age increased, instead decreased, during 1990–2010 indicating that transition in birth planning reversed in these states/Union Territories. In two states/Union Territories, the proportion of births to women below 20 years of age remained virtually unchanged so that transition in birth planning contributed little to fertility transition. In fact, there are 20 states/Union Territories where the proportion of births to women below 20 years of age increased, instead decreased, during 2000–2010. Similarly, in eight states/Union Territories, decrease in the proportion of third and higher order births during 1990–2000 was at best marginal and in three states—Chhattisgarh, Bihar and Uttar Pradesh—the proportion of third and higher-order births actually increased during this period.

The pattern of fertility transition across the districts can be analysed in terms of proportionate cumulative distribution of districts for the index *f*, as well as the proportionate cumulative distribution of the proportion of births to women below 20 years of age and the proportionate cumulative distribution of the proportion of third and higher-order births. The proportionate cumulative frequency curve of districts for the index *f* is presented in Fig. 4.1, whereas the proportionate cumulative frequency curve of districts for the proportion of births to women below 20 years of age and the proportion of third and higher-order births are presented in Figs. 4.2 and 4.3,

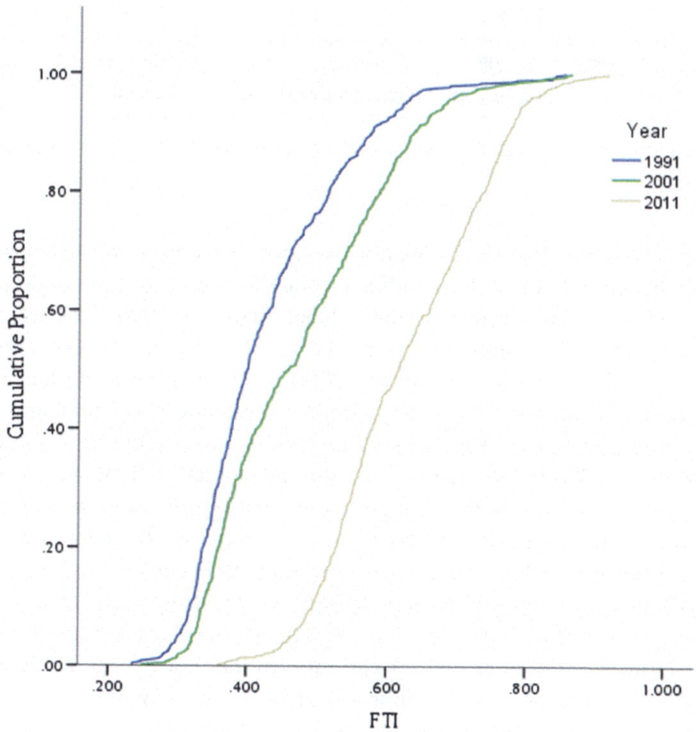

Fig. 4.1 Cumulative proportionate distribution of fertility transition index across districts

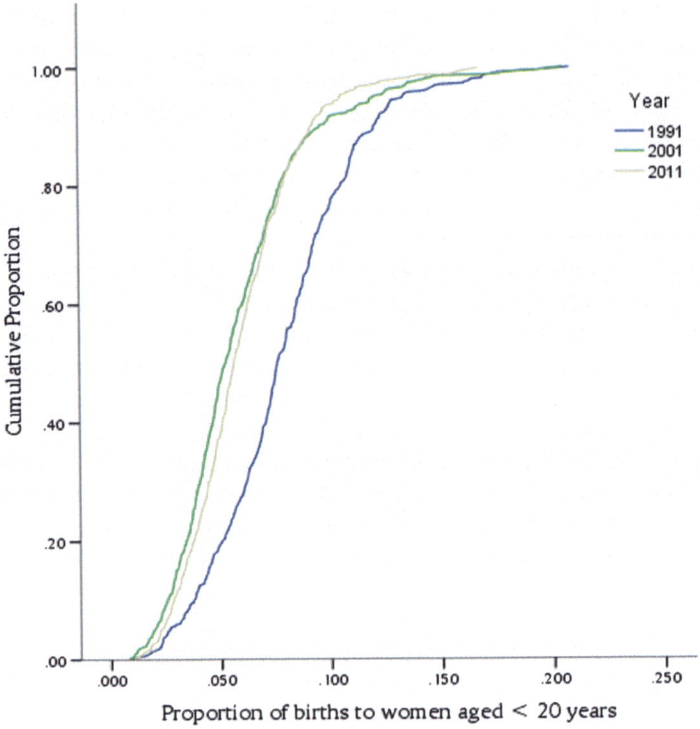

Fig. 4.2 Cumulative proportionate distribution of the proportion of births to women aged <20 years across districts

respectively. The horizontal shift towards the right in the proportionate cumulative frequency distribution of the index *f* indicates that the index *f*, in general, improved in all districts with time. This shift was relatively small during 1990–2000 as compared to that during 2000–2010 indicating that the fertility transition was slow during 1990–2000 but gained momentum during 2000–2010. Similar trend has also been observed at national and state levels also. Similarly, transition in birth planning nearly stagnated during 2000–2010 (Fig. 4.2). In contrast, transition in birth limitation was slow during 1990–2000 but gained momentum during 2000–2010 (Fig. 4.3).

It is difficult to decompose the change in the fertility transition index *f* over time into the change in the proportion of births to women below 20 years of age and the change in the proportion of third and higher-order births because of the changes in the boundaries of a large number of districts of between the 1991 population census and the 2011 population census as reflected through the increase in the number of districts since 1991. However, the inter-district variation in the index *f* can be decomposed into inter-district variation in the proportion of births to women below 20 years of age and inter-district variation in the proportion of third and higher-order births. It can be shown that

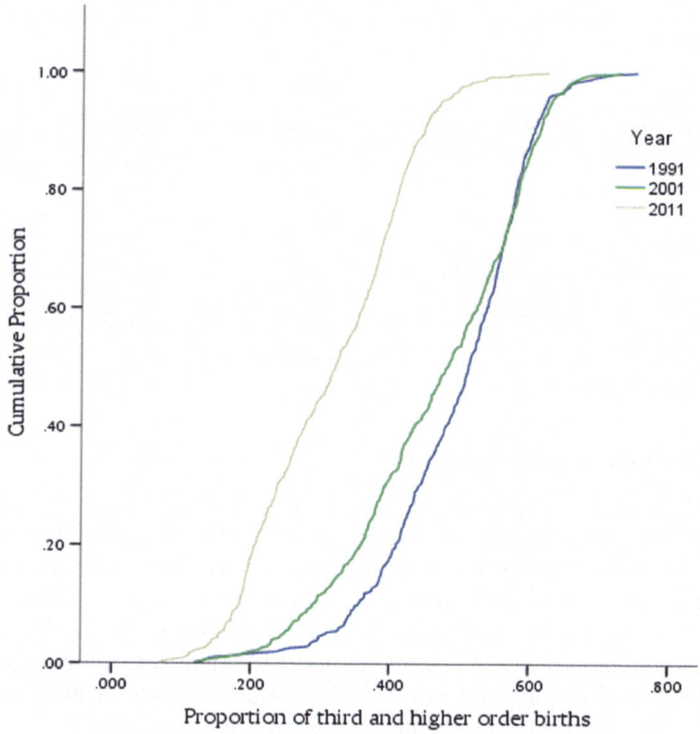

Fig. 4.3 Cumulative proportionate distribution of third and higher-order births across districts

$$var(f) = \left[var(b_{i<20}) + cov\left(b_{i<20}b_{j<3}\right)\right] + \left[var(b_{i<20}) + cov\left(b_{i<20}b_{j<3}\right)\right]$$

where *var* stands for variance and *cov* for covariance. In other words, the variation in the index f across the districts can be decomposed into two parts; one attributed to the variation in the birth planning dimension, while the other to the variation in the birth limitation dimension of fertility transition. This decomposition helps in analysing the relative importance of the two dimensions of fertility transition in explaining the variation in fertility transition across districts.

Results of the variance decomposition analysis are presented in Table 4.6. Nearly all the inter-district variation in the fertility transition index f is attributed to the inter-district variation in the proportion of third and higher-order births, either currently or in the past. Inter-district variation in the proportion of births to women below 20 years of age constitutes an insignificant proportion of the inter-district variation in the fertility transition index f. This implies that fertility transition in the districts of the country, during 1990–2010, was driven largely by the transition in the birth limitation dimension of fertility transition and not the dimension of birth planning. Moreover, the simple zero-order correlation coefficient between the two dimensions of fertility transition is found to be very close to zero which shows that the inter-district variation in the proportion of births to women below 20 years of age was largely independent of the inter-district variation in the proportion of third and higher-order births.

Table 4.6 Decomposition of the inter-district variance in the fertility transition index *f*, 1990–2011

Particulars	Year		
	1990	2000	2010
Variance (f)	0.01106	0.01455	0.01200
Variance ($b_i < 20$)	0.00107	0.00097	0.00069
Variance ($b_j < 3$)	0.01021	0.01519	0.01111
Covariance ($b_i < 20$, $b_j < 3$)	−0.00011	−0.00081	0.00010
Variance (f) accounted by variance ($b_i < 20$)	8.71000	1.13000	6.60000
Variance (f) accounted by variance ($b_j < 3$)	91.29000	98.87000	93.40000
Number of districts	452	593	640

Source Author's calculations

4.5 Classification of Districts

The inter-district variation in the index f suggests that there are district-specific factors that influence fertility transition. We have applied the segmentation or classification approach to identify characteristics of districts which are at different stages of fertility transition. The segmentation or classification approach is different from the regression approach that is commonly used for analysing marginal effects of independent or explanatory variables on the dependent variable. In the most general term, the segmentation or the classification approach is based on a set of *if–then* logical conditions that permit splitting districts into mutually exclusive groups in terms of the stage of fertility transition. There are different methods available for carrying out segmentation analysis. We have used the decision tree procedure and applied the classification and regression tree technique (CRT) to create a tree-based segmentation model (IBM Corporation 2012). CRT is a non-parametric recursive partitioning method (Brieman et al. 1984). It branches districts into nodes on the basis of selected explanatory variables to achieve within-cluster homogeneity. A node which cannot be split further is termed as a terminal node or the 'leaf'. A leaf in which all districts are at the same stage of fertility transition is called a homogenous or 'pure' node. Lack of homogeneity within a cluster is measured in terms of least-squared deviation which is computed as within-node variance (IBM Corporation 2012). The tree-growing process is continued until either the pure node is reached or the prescribed stopping criterion is met (Ambalavanan et al. 2006; Lemon et al. 2003). CART has a number of advantages for the purpose of segmentation. It makes no assumption about the variables used for segmentation. They may be a mix of categorical, interval, and scale variables. Results of CART are also not affected by the quality of data such as the presence of outliers, collinearity among the explanatory variables, heteroscedasticity, or distributional error structures that normally affect parametric procedures like regression.

In the present case, the dependent variable is the index f which is a scale variable. Initially, nine independent variables were used for segmentation analysis: (1)

degree of urbanisation measured in terms of the proportion of the urban population to the population of the district; (2) proportion of Scheduled Castes population; (3) proportion of Scheduled Tribes population; (4) literacy rate measured in terms of the proportion of the population aged at least 7 years who can read and write with understanding; (5) female literacy rate measured in terms of the proportion of females aged at least 7 years who can read and write with understanding; (6) child sex ratio measured as female/male ratio of the population aged 0–6 years; (7) sex ratio of the population aged 7 years and above measured as female/male ratio of the population at least 7 years of age; (8) population engaged in non-agricultural productive activities measured as the proportion of workers other than cultivators and agricultural labourers; and (9) sex ratio of the population engaged in non-agricultural productive activities. However, in order to minimise the risk of wrong segmentation, the number of independent variables was subsequently reduced to three: (1) female literacy rate; (2) sex ratio of the population aged 7 years and above; and (3) engagement in non-agricultural productive activities.

Results of the segmentation analysis are presented in Table 4.7 and the classification tree is depicted in Fig. 4.4. The 640 districts of the country as they existed at the 2011 population census can be branched into six terminal nodes or 'leaves' in terms of the stage of fertility transition. The first node or cluster (node 3) comprises 206 (32%) districts. These districts are characterised by very low female literacy rate (less than or equal to 61%) and sex ratio of population aged 7 years and above highly unfavourable to females (less than or equal to 977 females per 1000 males aged at least 7 years). Fertility in these districts, on average, is at early stage of transition (average $f = 0.552$). The average proportion of third and higher-order births in districts of this cluster was 39%, while average proportion of births to women below 20 years of age was around 6%.

The next cluster (node 4) comprises districts where female literacy is very low but the gender balance is comparatively favourable to females (more than 977 females per 1000 males aged at least 7 years). There are 72 (11.2%) districts in this cluster and fertility in these districts is at the middle stage of transition (average $f = 0.609$). The third cluster (node 7) comprises districts where female literacy is high (more than 61%) but the gender balance is highly unfavourable to females (less than or equal to 982 females for 1000 males aged at least 7 years) and the proportion of the population engaged in non-agricultural pursuits is low (less than 18%). There are 129 (20.2%) districts in this cluster and fertility, in these districts, on average, is also at the middle of transition (average $f = 0.613$). The fourth cluster (node 8) comprises districts where female literacy rate is at least 61% but the gender balance is unfavourable to females (less than or equal to 982 females per 1000 males aged 7 years and above) but more than 18% of the population is engaged in non-agricultural pursuits. There are 132 districts (20.6%) in this cluster and fertility in districts of this cluster is also at the middle of transition (average $f = 0.674$). The fifth cluster (node 9) comprises 50 (8%) districts. In these districts, female literacy rate is at least 61% and the gender balance is comparatively favourable to females but the proportion of population engaged in non-agricultural pursuits is less than 18%. Fertility in these districts is at advance stage of transition (average $f = 0.704$). Finally, the sixth and the last cluster

Table 4.7 Classification of districts by population characteristics and stage of fertility transition

Node	Population characteristics			Districts		FTI		Proportion of women <20 years		Proportion of third and higher-order births		Remarks
	Female literacy rate	Population sex ratio	Population engaged in non-agricultural activities	Number	Percent	Mean	SD	Mean	SD	Mean	SD	
0	All	All	All	640	100	0.626	0.110	0.059	0.026	0.315	0.105	
1	≤61	All	All	278	43.4	0.567	0.080	0.063	0.023	0.370	0.086	
2	>61	All	All	362	56.6	0.671	0.108	0.057	0.028	0.272	0.100	
3	≤61	≤968	All	206	32.2	0.552	0.073	0.062	0.024	0.387	0.077	Terminal
4	≤61	>968	All	72	11.2	0.609	0.084	0.067	0.021	0.323	0.091	Terminal
5	>61	≤982	All	261	40.8	0.644	0.102	0.060	0.029	0.300	0.096	
6	>61	>982	All	101	15.8	0.740	0.090	0.049	0.024	0.211	0.081	
7	>61	≤982	≤18	129	20.2	0.613	0.091	0.065	0.027	0.323	0.090	Terminal
8	>61	≤982	>18	132	20.6	0.674	0.103	0.055	0.030	0.271	0.095	Terminal
9	>61	>982	≤18	50	7.8	0.704	0.074	0.052	0.025	0.244	0.069	Terminal
10	>61	>982	>18	51	8.0	0.776	0.090	0.045	0.022	0.179	0.081	Terminal

Source Author's calculations

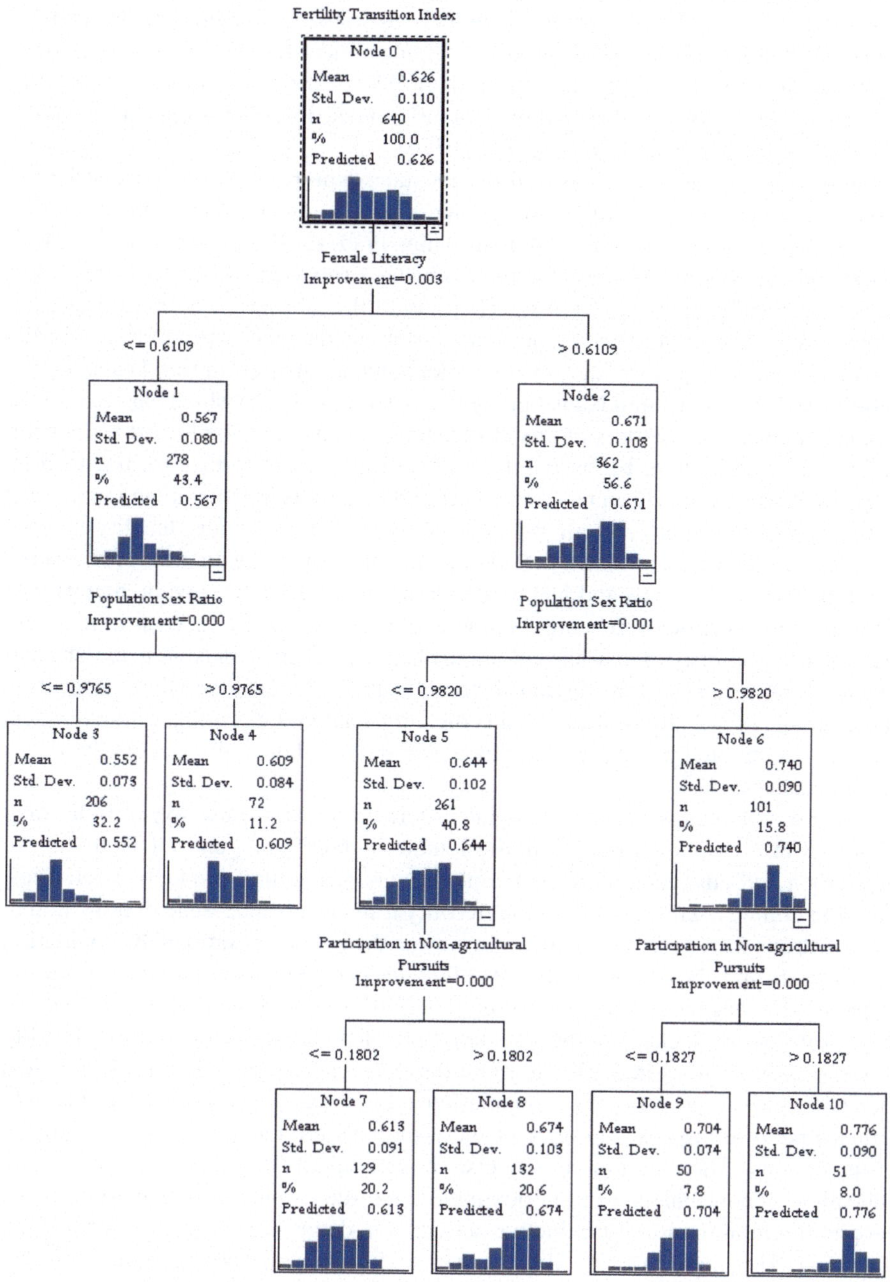

Fig. 4.4 Classification of districts by fertility transition index

(node 10) comprises 51 (8%) districts. In these districts, female literacy rate is more than 61%, gender balance is favourable to females (more than 982 females per 1000 males at least 7 years of age) and more than 18% of the population is engaged in non-agricultural productive activities. Fertility, in districts of this cluster, is also at advance stage of transition (average $f = 0.776$).

Table 4.8 shows how districts of different states/Union Territories are distributed across the six clusters identified through the segmentation analysis. Majority of districts in Jammu and Kashmir, Rajasthan, Uttar Pradesh, Bihar, Arunachal Pradesh Jharkhand and Madhya Pradesh belong to cluster 1 (node 3) which, on average, has the lowest fertility transition index f. Districts of this cluster have, on average, the highest proportion of third and higher order births and the highest proportion of births to women below 20 years of age. On the other hand, all districts in Kerala and almost two-thirds districts in Tamil Nadu belong to cluster 6 (node 10) which has the highest index f, on average. Districts of this cluster have, on average, the lowest proportion of third and higher-order births and the lowest proportion of births to women below 20 years of age. Similarly, in more than half of the districts in Uttarakhand, one-third districts in Chhattisgarh and almost 30% districts in Tamil Nadu, fertility is at an advanced stage of transition. On the other hand, majority of the districts in Himachal Pradesh, Punjab, Haryana and Delhi belong to cluster 4 (node 8) while majority of districts in the north-eastern states belong to cluster 3 (node 7). In West Bengal and Gujarat also, majority of districts belong to cluster 4, whereas in Assam, Odisha and Maharashtra, majority of the districts belong to cluster 3. Fertility in these districts is at the middle of transition. Karnataka is the only state in the country where districts are not concentrated in a particular cluster but are distributed almost equally across the six clusters.

Table 4.8 also suggests that regional pattern in fertility transition in India may be traced in the regional pattern in the female literacy rate, gender balance in the population and engagement of the people in non-agricultural productive activities. The data mining exercise suggests that fertility is at the advance stage of transition in those districts where female fertility rate is more than 61%, sex ratio of the population aged 7 years and above is more than 982 females per 1000 males and the proportion of population engaged in non-agricultural activities is more than 18%. Majority of these districts are located in the southern part of the country. In contrast, fertility is at the early stage of transition in those districts where female literacy rate is less than 61% and the sex ratio of the population aged 7 years and above is less than 977 females per 1000 males. Majority of these districts are located in the central part of the country. The data mining exercise thus highlights the importance of female education, gender balance in the population and the extent of participation of the people in non-agricultural productive activities in determining the stage of fertility transition. This means that fertility transition efforts should not be confined to the delivery of contraceptives only. Rather, they should be grounded in the broader human development perspective.

Table 4.8 Distribution of districts in different states/Union Territories by different terminal nodes)

India/State/Union territory	Node 3		Node 4		Node 7		Node 8		Node 9		Node 10		Total
	N	%	N	%	N	%	N	%	N	%	N	%	
India	206	32.2	72	11.3	129	20.2	132	20.6	50	7.8	51	8	640
Jammu and Kashmir	17	77.3	0	0	1	4.5	4	18.2	0	0	0	0	22
Himachal Pradesh	0	0	0	0	2	16.7	6	50.0	4	33.3	0	0	12
Punjab	2	10.0	0	0	3	15.0	15	75.0	0	0	0	0	20
Chandigarh	0	0	0	0	0	0	1	100.0	0	0	0	0	1
Uttarakhand	0	0	0	0	2	15.4	4	30.8	7	53.8	0	0	13
Haryana	6	28.6	0	0	5	23.8	10	47.6	0	0	0	0	21
Delhi	1	11.1	0	0	0	0	8	88.9	0	0	0	0	9
Rajasthan	25	75.8	5	15.2	1	3	2	6.1	0	0	0	0	33
Uttar Pradesh	49	69.0	5	7	11	15.5	6	8.5	0	0	0	0	71
Bihar	32	84.2	2	5.3	3	7.9	1	2.6	0	0	0	0	38
Sikkim	1	25.0	0	0	1	25	2	50.0	0	0	0	0	4
Arunachal Pradesh	9	56.3	3	18.8	1	6.3	2	12.5	1	6.3	0	0	16
Nagaland	0	0	0	0	8	72.7	3	27.3	0	0	0	0	11
Manipur	0	0	0	0	5	55.6	0	0	1	11.1	3	33.3	9
Mizoram	1	12.5	0	0	4	50.0	1	12.5	1	12.5	1	12.5	8
Tripura	0	0	0	0	2	50.0	1	25.0	1	25.0	0	0	4
Meghalaya	1	14.3	0	0	4	57.1	0	0	1	14.3	1	14.3	7
Assam	6	22.2	0	0	12	44.4	9	33.3	0	0	0	0	27
West Bengal	5	26.3	0	0	6	31.6	8	42.1	0	0	0	0	19

(continued)

Table 4.8 (continued)

India/State/Union territory	Node 3		Node 4		Node 7		Node 8		Node 9		Node 10		Total
	N	%	N	%	N	%	N	%	N	%	N	%	
Jharkhand	12	50.0	7	29.2	3	12.5	2	8.3	0	0	0	0	24
Odisha	1	3.3	12	40.0	10	33.3	5	16.7	2	6.7	0	0	30
Chhattisgarh	2	11.1	9	50.0	1	5.6	0	0	6	33.3	0	0	18
Madhya Pradesh	29	58	5	10.0	11	22.0	4	8.0	1	2.0	0	0	50
Gujarat	3	11.5	1	3.8	7	26.9	12	46.2	2	7.7	1	3.8	26
Daman and Diu	0	0	0	0	0	0	2	100.0	0	0	0	0	2
Dadra and Nagar Haveli	0	0	1	100.0	0	0	0	0	0	0	0	0	1
Maharashtra	1	2.9	1	2.9	19	54.3	8	22.9	6	17.1	0	0	35
Andhra Pradesh	0	0	15	65.2	1	4.3	2	8.7	4	17.4	1	4.3	23
Karnataka	2	6.7	6	20.0	6	20.0	7	23.3	4	13.3	5	16.7	30
Goa	0	0	0	0	0	0	1	50.0	0	0	1	50.0	2
Lakshadweep	0	0	0	0	0	0	0	0	0	0	1	100	1
Kerala	0	0	0	0	0	0	0	0	0	0	14	100	14
Tamil Nadu	1	3.1	0	0	0	0	2	6.3	9	28.1	20	62.5	32
Puducherry	0	0	0	0	0	0	1	25.0	0	0	3	75.0	4
Andaman and Nicobar Islands	0	0	0	0	0	0	3	100.0	0	0	0	0	3

Source Author's calculations

4.6 Discussion

In the context of India's National Population Policy 2000, first and second-order births to women aged 20 years and above are the 'desired' births and the policy aims at maximising the proportion of these 'desired' births in the quest towards 'small family' and population stabilisation. The present chapter analyses India's progress during 1990–2010 in this context. This progress has two distinct phases. Progress during 1990–2000 has been slow and primarily driven by transition in birth planning. Progress during 2000–2010 has been rapid and driven primarily by transition in birth limitation. During 1990–2000 transition in birth limitation nearly stagnated whereas transition in birth planning nearly stagnated during 2000–2010. This stagnation has implications in terms of reducing the effect of population momentum on future population growth and in terms of improving the survival and health of women, especially, mothers and children. India is still to go a long way to achieve the 'small family' norm as the proportion of 'undesired' births—births to women below 20 years of age and third and higher-order births - still constitute almost 40% of the reported births in 2010.

The analysis also suggests that the fertility transition path followed by India, especially, during 2000–2010, has been totally at odds to what has been envisaged and aimed in the National Population Policy 2000 and reiterated in the National Rural Health Mission (Government of India 2005) and in X and XI Five-year Development Plans of the country (Government of India 2008, 2013b). The Government of India claims that there has been a paradigm shift in the organised family planning efforts towards birth planning rather than birth limitation (Pachauri 2014; Rao 2000). However, the present analysis shows little evidence of such a shift. Fertility transition efforts in India continue to be preoccupied with birth limitation to reduce the completed fertility. Such an approach may lead to achieving the replacement fertility but it contributes little towards the realisation of the goal of stable population by 2045. In the context of population stabilisation, it is imperative that couples limit their family to two children and, at the same time, childbearing is delayed and births are properly spaced so that there is no decrease in the mean age of childbearing which is necessary to minimise the impact of momentum on future population growth (Bongaarts 1994; Ryder 1980). A focus on birth planning is also necessary in the context of reduction in infant and child mortality and improvement in the health status of women so that the demand for 'bonus' children is reduced, thereby accelerating transition in birth limitation. There was some progress in this direction during the 1990s as is reflected through the decrease in proportion of births to women below 20 years of age, but the initiative appears to have lost during 2000–2010. Mathews et al. (2009) has estimated that adopting a 'later, longer and fewer' family planning strategy would lead India to a future population total which would be about 52 million less in 2050 than if the current birth limitation-based strategy is continued. There is, however, little evidence on any movement of fertility transition efforts in this direction despite a clear emphasis in the National Population Policy 2000.

The evidence of the stagnation or even reversal in transition in birth planning in India during 2000–2010 is also available from other sources. According to the Sample Registration System, about 29% births in India in 1991 had a birth interval of at least 3 years. This proportion increased to 42.5% in 2001 but decreased to 40.7% in 2013. Similarly, the proportion of births having a birth interval of less than 24 months was 38.5% in 1991, which decreased to 23.9% in 2001 but increased to almost 30% in 2013. These figures confirm that the priority given to birth planning during the 1990s was lost during the 2000s, probably and so obviously, because the priority given to birth planning resulted in the neglect of birth limitation in fertility transition efforts. This neglect of birth planning in fertility transition efforts is bound to cost India dear in the context of population stabilisation as shown by the analysis of Mathews et al. (2009).

Interestingly, the main driver of fertility transition has been found to be different during 1990–2000 than during 2000–2010. During the 1990–2000, fertility transition was driven primarily by transition in birth planning at the cost of near stagnation in transition in birth limitation. The scenario reversed during 2000–2010 when fertility transition was driven primarily by the transition in birth limitation while transition in birth planning stagnated. This complementarity in family planning efforts indicates lack of a comprehensive, multi-dimensional approach towards fertility transition in the country. Fertility transition efforts in India have largely been limited to the delivery of family planning services through the official family planning efforts, but the poor organisational efficiency and low administrative capacity of these efforts are well known (Chaurasia 1985; Chaurasia and Singh 2014). The present analysis again reiterates the oft-repeated need of going beyond family planning to accelerate and properly orient fertility transition and population stabilisation efforts in the country. It emphasises that the necessary threshold for hastening the pace of fertility transition should be traced in gender balance in the population, female education and participation in non-agricultural productive activities, as stipulated in the National Population Policy 2000.

One reason, probably and so obviously, behind the neglect of the birth planning dimension in fertility transition and population stabilisation efforts in India appears to be the absence of any effective system of measuring and monitoring transition in birth planning. The pre-occupation with achieving replacement fertility at the policy level has resulted in a focus on birth limitation at the grassroots level. Since TFR is directly related to the proportion of third and higher-order births, the focus of family planning in India has always been on preventing rather than planning births. This pre-occupation is amply reflected in terms of monitoring the proportion of third and higher-order births in the official health management information system, which has been put in place under the National Rural Health Mission. Interestingly, there is no place for the proportion of births to women below 20 years of age in this system.

4.7 Conclusions

The present analysis highlights the limitations of fertility transition efforts in India in the context of National Population Policy 2000. Fertility transition efforts in the country remain confined to birth limitation despite all talks of the paradigm shift at the policy level. There is evidence of transition in birth planning in the country during the 1990s but this transition either stagnated or reversed during the 2000s. The current pattern of fertility transition in India contributes little towards minimising the effect of momentum on future population growth. In 11 major states of the country, the replacement fertility has already been achieved. In these states, the population will continue to increase in the coming years because of the momentum for growth built in the population. The effect of the momentum can be delayed through increasing the mean age of childbearing but there appears little initiative in this direction. It is obvious that birth limitation efforts in these states are hardly relevant in the quest for population stabilisation. Moreover, the current fertility transition pattern contributes, at best marginally, to the reduction in maternal and child mortality and improvement in maternal and child health in terms of the delay in the first birth and increase in spacing between successive births. However, the strategy of shifting the focus of family planning efforts from birth limitation to birth planning initiated during the 1990s could not be sustained primarily because family planning efforts lacked the organisational efficiency and the administrative capacity necessary for such a shift. At the same time, the present analysis again reiterates the need of integrating fertility transition efforts with broader human development activities. This observation is also not new but there appears little indication of such an integration. It is well known that a certain threshold of human development is necessary for the effectiveness of fertility transition efforts in modifying the reproductive behaviour of couples in favour of planning births rather than preventing births. There, however, appears little progress in this direction. Lack of such integration, probably and so obviously, appears to be a reason why fertility transition in India remains confined to preventing births.

References

Ambalavanan N, Baibergenova A, Carlo WA, Saigal S, Scmidt B, Thorpe KE (2006) Early prediction of poor outcome in extremely low birth weight infants by classification tree analysis. J Pediatr 148(4):438.e1–444.e1

Balasubramanian S (1972) An analysis of declining fertility in greater Bombay through birth order statistics. In: Paper presented at the all India seminar on family planning problems in India. International Institute for Population Studies, Bombay

Bertrand JT, Magnani RJ, Rutenberg N (1994) Handbook of indicators for family planning program evaluation. Chapel Hill, University of North Carolina at Chapel Hill, Caroline Population Center. The Evaluation Project

Bongaarts J (1994) Population policy options in the developing world. New York, The Population Council. Research Division Working Paper No. 59

Brieman L, Friedman J, Olshen RA, Stone CJ (1984) Classification and regression trees, Wadsworth

CATALYST Consortium (2009) Systematic literature review and meta-analyses on birth Spacing: how birth spacing relates to infant and child mortality, maternal and perinatal health, and maternal and child nutrition outcomes. Pathfinder International, Washington DC
Chidambaram VC (1965) Changes resulting from family planning programme in the indices based on birth order statistics. In: Paper submitted to the seminar on new approaches in the use of mathematical models in demographic research. Demographic Research Centre, Mumbai
Chaurasia AR (1985) Organizational aspects of family welfare programme in India. Indian J Community Med 10(3&4):181–190
Chaurasia AR (2013a) Fertility transition in India: evidence from DLHS 2007–2008. J Fam Welf 59(1):12–32
Chaurasia AR (2013b) Fertility transition in districts of India. Evidence from district level household and facility survey 2007–2008. In: Paper presented at XXVII IUSSP international population conference, Busan. International Union for the Scientific Study of Population
Chaurasia AR (2016) Demographic drivers of future population growth in India. Demogr India 44(1&2):83–95
Chaurasia AR, Gulati SC (2008) India: The State of Population 2007. Oxford University Press, New Delhi
Chaurasia AR, Singh R (2014) Forty years of planned family planning efforts in India. J Fam Welf 60(2):1–16
Cleland JG, Sathar ZA (1984) The effect of birth spacing on childhood mortality in Pakistan. Popul Stud 38(3):401–418
Conde-Agudelo A, Belizán M (2000) Maternal morbidity and mortality associated with interpregnancy interval: cross sectional study. BMJ 321(7271):1255–1259
Conde-Acqudelo A, Rosas-Bermudez A, Castano F, Norton MH (2012) Effects of birth spacing on maternal, perinatal, infant and child health: a systematic review of causal mechanisms. Stud Fam Plann 43(2):93–114
DaVanzo J, Razzaque A, Rahman M, Hale L (2004) The effects of birth spacing on infant and child mortality, pregnancy outcomes, and maternal morbidity and mortality in Matlab, Bangladesh. Rand Corporation. WR-198
Espenshade TJ, Guzmann JC, Westoff CF (2003) The surprising global variation in replacement fertility. Popul Res Policy Rev 22(5–6):575583
Fotso JC, Cleland J, Mberu B, Mutua M, Elungata P (2013) Birth spacing and child mortality: an analysis of prospective data from Nairobi urban health and demographic surveillance system. J Biosoc Sci 45(6):779–798
Frejka T (1982) Momentum. In: Ross JA (ed) International Encyclopedia of Population, vol 2. Free Press, New York
Government of India (1997) Census of India, 1991: population projections for India and States, 1996–2016. Registrar General and Census Commissioner, New Delhi
Government of India (2000) National population policy 2000. Ministry of Health and Family Welfare, New Delhi
Government of India (2005) National rural health mission: 2005–2012. Ministry of Health and Family Welfare, New Delhi
Government of India (2007) Population projections for India and States 2001–2026. Registrar General and Census Commissioner of India, New Delhi
Government of India (2008) Eleventh five-year plan (2007–2012) social sector, vol II. Planning Commission, New Delhi
Government of India (2013a) Twelfth five-year plan (2012–2017) social sectors, vol III. Planning Commission, New Delhi
Government of India (2013b) Sample registration system statistical report 2011. Ministry of Home Affairs, Office of the Registrar General, New Delhi
Government of India (2014a) Sample registration system statistical report 2013. Office of the Registrar General and Census Commissioner, New Delhi

Government of India (2014b) Census of India 2011. Report on post enumeration survey. Registrar General and Census Commissioner, New Delhi, India

Government of India (2017) National health policy 2017. Ministry of Health and Family Welfare, New Delhi

Government of India (no date) Sample Registration System statistical report 2017. Registrar General and Census Commissioner, India, New Delhi

Hobcraft JN, McDonald JW, Rutstein SO (1985) Demographic determinants of infant and early child mortality: a comparative analysis. Popul Stud 39(3):363–385

IBM Corporation (2012) IBM SPSS decision trees. IBM Corporation

Jasilioniene A, Jdanov DA, Sobotka T, Andreev EM, Zeman K, Shkolnikov VM (2012) Methods protocol for the human fertility database. MPIDR

Keyfitz N (1971) On the momentum of population growth. Demography 8(1):7180

Lemon SC, Roy J, Clark MA, Friedmann PD, Rakowski W (2003) Classification and regression tree analysis in public health: methodological review and comparison with logistic regression. Ann Behav Med 26(3):171–181

Mathews Z, Padmadas SS, Hutter I, McEachran J, Brown JJ (2009) Does early child-bearing and a sterilization-focussed family planning programme in India fuel population growth? Demogr Res 20(28):693–720

Merrick TW (1986) World population in transition. Population Bulletin 41(2)

Moreland S, Talbird S (2006) Achieving the Millennium Development Goals: the contribution of fulfilling the unmet need for family planning. Futures Group/POLICY Project and USAID

Pachauri S (2014) Priority strategies for India's family planning programme. India J Med Res 140(Supplement):137–146

Prasartakul P, Porapakkham Y, Sittitrai W (1987) Report on birth order distribution as a family planning programme evaluation indicator. Mahidol University, Institute for Population and Social Research, Bangkok

Rao M (2000) Family planning programme. Paradigm shift in strategy. Econ Polit Wkly 35:4317–4322

Ravenholt RT, Frederiksen H (1968) Numeration analysis of fertility patterns. Public Health Rep 83(6):449–457

Reynolds J (1972) Evaluation of family planning programme performance: a critical review. Demography 9(1)

Ryder NB (1980) Components of temporal variations in American fertility. In: Hirons RW (ed) Demographic patterns in developed societies. Taylor and Francis, London

Ryder NB (1982) Progressive fertility analysis. International Statistical Institute, Voorburg. World Fertility Survey Technical Bulletin No. 8

Singh P (2002) Trends in fertility mortality, nutrition and health indicators. In: Background paper prepared for the Planning Commission of India. http://planningcommission.nic.in/reports/genrep/bkpap2020/23_bg2020.pdf

Sivakumar MN (2000) Age at first birth and fertility of women in Kerala: a cohort analysis. Man India 80(3–4):237–287

Smith R, Ashford L, Gribble J, Clifton D (2009) Family planning saves lives. Population Reference Bureau

Srinivasan K, Freymann J (1990) Need for reorientation of family planning programme strategies in developing countries. A case for birth-based approach. In: Srinivasan K, Pathak KB (eds) Dynamics of population and family welfare, 1989. Himalaya Publishing House, Mumbai

Srinivasan K, Saxena PC, Pandey A (1992) Birth order and birth interval statistics. International Institute for Population Sciences, Mumbai [Unpublished]

United Nations (1997) Family building and family planning evaluation. Department of Economic and Social Affairs, Population Division, New York

United Nations (2017) World population prospects: the 2017 revision, vol I. Comprehensive tables. Department of Economic and Social Affairs, Population Division, New York. ST/ESA/SER.A/379

UNDP (United Nations Development Programme) (2003) Achieving the millennium development goals: population and reproductive health as critical determinants. United Nations Development Programme, New York. Population and Development Strategies Series

WHO (World Health Organization) (2005) Report of a WHO technical consultation on birth spacing. WHO, Geneva

Young AT, Higgins HJ, Levy D (2008) Sigma convergence versus beta convergence. Evidence from US county-level data. J Money Credit Banking 40(5):1083–1093

Chapter 5
Child Survival, 1971–2015

Abstract This chapter analyses child survival progress in India during the 40 years between 1971 and 1975 through 2011–2015 in terms of the improvement in the survival probability during the first 18 years of life. The chapter highlights the unsatisfactory progress towards securing the life of Indian children during the 40 years under reference. A disturbing finding of the present analysis is that the child survival progress in the country appears to have slowed down after 1991–1995 which appears to be the reason behind the country missing the MDG 4 of the United Nations Millennium Development Agenda. From the global perspective, an accelerated improvement in child survival probability in India is the need of the time so as to achieve the targets set under United Nations' 2030 Sustainable Development Agenda.

5.1 Introduction

Concerns about securing the life of the Indian children date back to 1946 when the First Health Survey and Development Committee, constituted during the colonial rule, recommended that measures directed towards reducing sickness and mortality in children must be given the highest priority in the programme for health development in the country (Government of India 1946). After independence, improving the survival of children has been one of the priority development agenda in all Five-year Development Plans of the country. In 1974, the Government of India announced the first National Policy on Children which declared children as country's 'supremely important asset' and emphasised that the programmes for children should be given prominent place in national plans for the development of human resources, so that children could grow up to become robust citizens, physically fit, mentally alert and morally healthy, endowed with the skills and motivations provided by the society (Government of India 1974). In 1975, the Integrated Child Development Scheme was launched, which is now the world's largest community-based outreach programme for early childhood development which has implications for securing the life of children (Government of India 1975).

Concerted efforts towards improving child survival in India could, however, be started only in 1978 when the Expanded Programme of Immunisation and National

© Springer Nature Singapore Pte Ltd. 2020
A. R. Chaurasia, *Population and Sustainable Development in India*,
https://doi.org/10.1007/978-981-32-9212-3_5

Programme for the Control of Diarrhoeal Diseases were launched (Sokhey et al. 1989; World Health Organization 1986; Tyagi 1983). The Expanded Programme of Immunisation graduated into the Universal Immunisation Programme in 1985 and was given the status of National Technology Mission in 1986 (Government of India 1988). In 1989, the National Acute Respiratory Infections Control Programme was launched and, in 1992, different vertical programmes and activities directed towards health of mothers and children were subsumed into the Child Survival and Safe Motherhood Programme (Government of India 1992) which was expanded into the Reproductive and Child Health Programme in 1997 (Government of India 1997). The Reproductive and Child Health Programme has been the lead programme of the National Rural Health Mission (2005–2012) launched in 2005 (Government of India 2005), which has since been expanded into the National Health Mission.

There have been other initiatives also to address child survival and health-related issues in India. These include pulse polio initiative launched in 1995, the constitution of a Technical Committee on Child Health and the Immunisation in 2000 and formation of a National Technical Advisory Group on Immunisation in 2001 (Government of India 2003). In 2013, the Government of India announced the new National Policy on Children which affirmed that the right to life, survival, health and nutrition was an inalienable right of every child (Government of India 2013). The recently announced India's National Health Policy 2017 aims at reducing the under-five mortality rate to 23 under-five deaths for every 1000 live births by 2025 through reducing the infant mortality rate to 28 infant deaths for every 1000 live births by 2019 and neonatal mortality rate to 16 neonatal deaths for every 1000 live births by 2025 (Government of India 2017a). The policy emphasises addressing social determinants of child survival and health through developmental action in all sectors.

How have India's child survival efforts contributed towards improving the survival chances of Indian children? The available evidence suggests that the survival probability during the first 5 years of life has improved substantially during the last 40 years. This improvement is often cited as the evidence of the success of the child survival movement in India. However, the pace of improvement has been slower than expected, so child survival probability in India remains low by international standards. The estimates prepared by the United Nations Inter-agency Group for Child Mortality Estimation (UNICEF 2015) suggest that the under-five mortality rate in India decreased by 62% between 1990 and 2015 so that the Millennium Development Goal 4 could not be achieved. It is estimated that India alone accounted for more than 20% (about 1.2 million) of about 5.9 million deaths of children below 5 years of age in the world in 2015. It is also argued that about 57% of the deaths in children below 5 years of age in India can be prevented through achieving high coverage of basic public health and nutrition interventions (Jones et al. 2006).

Nearly all studies on child survival progress in India have been limited to the survival in either the first year of life or the first 5 years of life (Puffer 1985; Visaria 1985; Jain and Visaria 1988; Tilak 1991; Pandey et al. 1998; Measham et al. 1999; Cleason et al. 2000; NIMS and UNICEF 2012; Kuntla et al. 2014; NIPCCD 2014). The National Policy on Children 2013 (Government of India 2013), however, classifies a person as a child if the person is below 18 years of age. The United Nations

Convention on the Rights of the Child also recognises a person as a child if the person is below 18 years of age (United Nations 1989). The Constitution of India (Government of India 2015) and the Child Labour (Prohibition and Regulation) Amendment Act 2016 (Government of India 2016a), on the other hand, consider people below 15 years of age as children. There is, however, little analysis of the survival prospects of Indian children aged 5–14 years or 5–17 years. In the context of the National Policy on Children 2013, it is imperative that the progress towards child survival should be analysed in terms of the probability of survival during the childhood or up to the 18th birthday in addition to the probability of survival up to the 5th birthday.

Analysing child survival progress in terms of the probability of survival up to the 18th birthday is also important in the context of the concept of the continuum of care that has been highlighted as a core principle of programmes for improving maternal, new born and child health, and as a means to reduce the burden of unwanted maternal and child deaths (WHO 2005; Tinker et al. 2005; de Graft-Johnson et al. 2006). The concept of the continuum of care aims at avoiding dichotomies, between either mothers and children, places of service delivery or single health issues (Kerber et al. 2007; Lawn et al. 2006; OECD 2005). In the context of child survival, the concept of the continuum of care ensures that all new born are able to survive the childhood period or the first 18 years of life and thrive (Gill et al. 2007).

This chapter analyses child survival progress in India during the 40 years between 1971–1975 through 2011–2015 in terms of the improvement in the survival probability during the first 18 years of life, including the progress during the first 5 years of life. From the perspective of continuum of care, the childhood period can be divided into four distinct phases: (1) infancy (<1 year); (2) preschool age (1–5 years); (3) school age (5–10 years); and (4) adolescence (10–18 years). This means that the survival through the period of childhood depends upon the survival during infancy, pre-school age school age and adolescence. We use a decomposition approach to estimate the relative contribution of the improvement in survival probability during different ages of the childhood period to the improvement in the survival probability during childhood.

The chapter is organised as follows: The next section describes the methodology adopted for estimating the probability of survival up to the 18th birthday from the official abridged life tables prepared and published by the Government of India and for decomposing the change in the probability of survival up to 18th birthday and up to 5th birthday over time. The chapter uses a decomposition methodology to analyse the contribution of the change in the probability of survival during infancy, pre-school age school age and adolescence to the improvement in the survival probability during childhood. Section three describes the data source. The chapter is based on the data available through India's official Sample Registration System. Section four discusses levels and trends and differentials in the probability of survival of a new born up to the 18th birthday. Section four analyses the improvement in survival probability at the national level, whereas section five decomposes the change in the survival probability during childhood into the change in the survival probability during infancy, pre-school age school age and adolescence. Section six analyses inter-state variations in the progress towards child survival while the last section summarises main findings of the analysis and discusses future prospects of improvement in the child survival probability in the country, especially in the context of the National Health Policy 2017 and the 2030 Sustainable Development Agenda of the United Nations.

5.2 Methodology

The key variable used in the present analysis is the probability of survival during childhood or, more specifically, the probability of survival from birth to the 18th birthday ($_{18}p_0$). It is well known that the probability of survival up to the 18th birthday ($_{18}p_0$) can be obtained as

$$_{18}p_0 = \prod_{i=1}^{18} {_i}p_{i-1} \tag{5.1}$$

where $p = 1 - q$; and q denotes the probability of death at a particular age. More specifically, if the childhood period (0–18 years) is divided into four age segments 0–1 year; 1–5 years; 5–10 years; and 10–18 completed years, then Eq. (5.1) can be written as:

$$_{18}p_0 = {_1}p_0 * {_4}p_1 * {_5}p_5 * {_8}p_{10} \tag{5.2}$$

Equation (5.2) implies that the change in $_{18}p_0$ over two points in time can be written as:

$$\nabla_{18}p_0 = {_{18}}p_0^2 - {_{18}}p_0^1$$

or

$$\nabla_{18}p_0 = {_1}p_0^2 * {_4}p_1^2 * {_5}p_5^2 * {_8}p_{10}^2 - {_1}p_0^1 * {_4}p_1^1 * {_5}p_5^1 * {_8}p_{10}^1 \tag{5.3}$$

Now, following Kim and Strobino (1984) and Das Gupta (1993) and applying the Goldfield's rule of 'allocating interactions to different individual factors on the basis of the principle of equal distribution of all variables involved in each interaction' (Durand 1948), the change in $_{18}p_0$ can be decomposed as:

$$\nabla_{18}p_0 = \partial_1 p_0 + \partial_4 p_1 + \partial_5 p_5 + \partial_8 p_{10} \tag{5.4}$$

where

$$\partial_1 p_0 = A + \frac{AB + AC + AD}{2} + \frac{ABC + ABD + ACD}{3} + \frac{ABCD}{4} \tag{5.5}$$

$$\partial_4 p_1 = B + \frac{BA + BC + BD}{2} + \frac{ABC + ABD + BCD}{3} + \frac{ABCD}{4} \tag{5.6}$$

$$\partial_5 p_5 = C + \frac{AC + BC + CD}{2} + \frac{ABC + ACD + bcd}{3} + \frac{ABCD}{4} \tag{5.7}$$

$$\partial_8 p_{10} = D + \frac{AD + BD + CD}{2} + \frac{ABD + ACD + BCD}{3} + \frac{ABCD}{4} \tag{5.8}$$

and

$$A = \left({}_1p_0^2 - {}_1p_0^1\right) * {}_4p_1^1 * {}_5p_5^1 * {}_8p_{10}^1 \tag{5.9}$$

$$B = {}_1p_0^1 * ({}_4p_1^1 - {}_4p_1^1) * {}_5p_5^1 * {}_8p_{10}^1 \tag{5.10}$$

$$C = {}_1p_0^1 * {}_4p_4^1 * ({}_5p_5^2 - {}_1p_0^1) * {}_8p_{10}^1 \tag{5.11}$$

$$D = {}_1p_0^1 * {}_4p_4^1 * {}_1p_0^1) * ({}_8p_{10}^2 - {}_8p_{10}^1) \tag{5.12}$$

$$AB = \left({}_1p_0^2 - {}_1p_0^1\right) * ({}_4p_1^2 - {}_4p_1^1) * {}_5p_5^1 * {}_8p_{10}^1 \tag{5.13}$$

$$AC = \left({}_1p_0^2 - {}_1p_0^1\right) * {}_4p_1^1 * ({}_5p_5^2 - {}_5p_5^1) * {}_8p_{10}^1 \tag{5.14}$$

$$AD = \left({}_1p_0^2 - {}_1p_0^1\right) * {}_4p_1^1 * {}_5p_5^1 * ({}_8p_{10}^2 - {}_8p_{10}^1) \tag{5.15}$$

$$BC = {}_1p_0^1 * ({}_4p_1^1 - {}_4p_1^1) * ({}_5p_5^2 - {}_5p_5^1) * {}_8p_{10}^1 \tag{5.16}$$

$$BD = {}_1p_0^1 * ({}_4p_1^1 - {}_4p_1^1) * {}_5p_5^1 * ({}_8p_{10}^2 - {}_8p_{10}^1) \tag{5.17}$$

$$CD = {}_1p_0^1 * {}_4p_4^1 * ({}_5p_5^2 - {}_1p_0^1) * ({}_8p_{10}^2 - {}_8p_{10}^1) \tag{5.18}$$

$$ABC = \left({}_1p_0^2 - {}_1p_0^1\right) * ({}_4p_1^2 - {}_4p_1^1) * ({}_5p_5^2 - {}_5p_5^1) * {}_8p_{10}^1 \tag{5.19}$$

$$ABD = \left({}_1p_0^2 - {}_1p_0^1\right) * ({}_4p_1^2 - {}_4p_1^1) * {}_5p_5^1 * ({}_8p_{10}^2 - {}_8p_{10}^1) \tag{5.20}$$

$$ACD = \left({}_1p_0^2 - {}_1p_0^1\right) * {}_4p_1^1 * ({}_5p_5^2 - {}_5p_5^1) * ({}_8p_{10}^2 - {}_8p_{10}^1) \tag{5.21}$$

$$BCD = {}_1p_0^1 * ({}_4p_1^1 - {}_4p_1^1) * ({}_5p_5^2 - {}_5p_5^1) * ({}_8p_{10}^2 - {}_8p_{10}^1) \tag{5.22}$$

$$ABCD = \left({}_1p_0^2 - {}_1p_0^1\right) * ({}_4p_1^2 - {}_4p_1^1) * ({}_5p_5^2 - {}_5p_5^1) * ({}_8p_{10}^2 - {}_8p_{10}^1) \tag{5.23}$$

Equation (5.4) analyses the change in the survival probability during the first 18 years of life in terms of the change in the survival probability during the first year of life, pre-school age, school age and adolescence. It is well known that the main causes of death in children of different ages are different. Deaths due to prematurity and low birth weight, pneumonia, birth asphyxia and birth trauma accounted for almost 63% of all infant deaths in India during 2010–2013 (Government of India no date). Deaths due to pneumonia, diarrhoeal diseases, injuries and other communicable diseases accounted for almost 64% of all deaths of children aged 1–5 years, whereas unintentional injuries other than motor vehicle accidents, diarrhoeal diseases and other infectious and parasitic diseases accounted for almost 63% of all deaths of children aged 5–15 years. In children aged 15 years and above, the main causes of death are intentional (suicides) and unintentional injuries. Obviously, the change in the survival probability in different age groups may be attributed to specific child survival interventions.

Similarly, the probability of survival in the first 5 years of life can be decomposed as:

$$\nabla_5 p_0 = \partial_1 p_0 + \partial_4 p_1 \tag{5.24}$$

where

$$\partial_1 p_0 = E + \frac{EF}{2} \tag{5.25}$$

$$\partial_4 p_1 = F + \frac{EF}{2} \tag{5.26}$$

and

$$E = \left({}_1p_0^2 - {}_1p_0^1\right) * {}_4p_1^1 \tag{5.27}$$

$$F = {}_1p_0^1 * \left({}_4p_1^1 - {}_4p_1^1\right) \tag{5.28}$$

$$EF = \left({}_1p_0^2 - {}_1p_0^1\right) * \left({}_4p_1^2 - {}_4p_1^1\right) \tag{5.29}$$

5.3 Data Source

The analysis is based on the official abridged life tables prepared by the Registrar General and Census Commissioner of India beginning 1971–1975 through 2011–2015 based on the Sample Registration System (Government of India 1984; 1985; 1989; 1994; 2004; 2008; 2017). The Sample Registration System is a dual-record system which provides annual estimates of birth rate, death rate and other indicators of fertility and mortality indicators at national and sub-national levels (Lal and Swamy 1977). The system was launched in 1964–1965 and became fully operational in

1969–1970. It is currently operational in 8779 sample units—4909 rural units and 3870 urban units—spread across all the states and Union Territories of the country, except the Ladakh region of Jammu and Kashmir. In 2016, the system covered more than 7.05 million population (Government of India 2016a).

The abridged life tables prepared by the Registrar General and Census Commissioner of India provide estimates of the probability of death for the conventional age groups: 0–1 year; 1–5 years; 5–10 years; 10–15 years; 15–20 years; and so on, but not for the childhood period or the age group 0–18 years. We have therefore derived unabridged life tables from the official abridged life tables by applying the UNABR routine of the MORTPAK-Lite software package of mortality analysis (United Nations 1988). The UNABR routine is based on the Heligman-Pollard model mortality schedule (Heligman and Pollard 1980) for expanding the abridged life table. Rogers and Gard (1981) have observed that UNABR routine gives 'correct' estimates of the parameters of the Heligman-Pollard mortality model and the aggregation bias that arise from using 5-year instead of single-year age groups in fitting the model is probably acceptably small in most instances, with the exception of the parameter E which measures the spread of 'accident' hump. On the basis of the single-year probability of death obtained through the application of the UNABR routine, we have calculated ${}_1p_0$, ${}_5p_0$, ${}_{10}p_0$ and ${}_{18}p_0$, then ${}_4p_1$, ${}_5p_5$ and ${}_8p_{10}$ have been calculated for the total population, as well as separately for males and females for different 5 years interval for the period 1971 through 2015. These survival probabilities constitute the database for the present analysis.

5.4 Improvement in Child Survival Probability

Estimates of the probability of survival up to the 18th birthday, ${}_{18}p_0$, along with the estimates of ${}_1p_0$, ${}_4p_1$, ${}_5p_0$, ${}_5p_5$ and ${}_8p_{10}$ for nine population sub-groups are presented in Appendix Table A.5.1, while the improvement in the child survival probability during the period 1971–1975 through 2011–2015 is presented in Table 5.1. During 1971–1975, only 756 out of every 1000 new born survived to their 18th birthday in the country. This number increased to 940 during 2011–2015. Similarly, 789 out of every 1000 new born survived to their 5th birthday during 1971–1975. This number increased to 949 during 2011–2015. The increase in ${}_{18}p_0$ and ${}_5p_0$ was the most rapid in rural females but the least rapid in urban males. In the rural areas, the increase in ${}_{18}p_0$ and ${}_5p_0$ was almost two times more than the increase in the urban areas. Similarly, increase in ${}_{18}p_0$ and ${}_5p_0$ was more rapid in female as compared to male children. As a result, the gender and residence gap in both ${}_{18}p_0$ and ${}_5p_0$ has narrowed substantially over time, although, male survival probability remains higher than the female survival probability and survival probability in the urban areas remains well above the survival probability in the rural areas.

Alternatively, out of every 1000 new born during 1971–1975, 244 failed to survive to their 18th birthday—132 died in the first year of life; 79 died during 1–4 years of life; 18 died in during 5–9 years of life; and 14 died during 10–17 years of life

Table 5.1 Improvement in child survival probability in India, 1971–2015

Population		Survival probability					
		$_{1}p_{0}$	$_{4}p_{1}$	$_{5}p_{0}$	$_{5}p_{5}$	$_{8}p_{10}$	$_{18}p_{0}$
Total		0.090	0.082	0.160	0.019	0.012	0.183
Total	Male	0.088	0.070	0.149	0.017	0.010	0.170
Total	Female	0.091	0.095	0.172	0.021	0.015	0.198
Rural		0.095	0.090	0.171	0.021	0.014	0.196
Rural	Male	0.093	0.079	0.160	0.019	0.011	0.181
Rural	Female	0.097	0.103	0.184	0.023	0.017	0.211
Urban		0.061	0.036	0.094	0.010	0.007	0.107
Urban	Male	0.059	0.025	0.081	0.009	0.006	0.094
Urban	Female	0.054	0.048	0.097	0.011	0.007	0.112

Source Author's calculations

(Fig. 5.1). By comparison, out of every 1000 new born during 2011–2015, 60 are likely to fail to survive to their 18th birthday—42 would have died in the first year of life; 8 are expected to die during 1–4 years of life; 4 are expected to die during their 5–9 years of life; and 6 are expected to die during their 10–17 years of life. Figure 5.1 also suggests that the decrease in the number of deaths during 1–4 years of life for every 1000 new born was quite rapid during the period 1971–1975 through 1991–1995 but the decrease appears to have slowed down considerably during the period 1991–1995 through 2011–2015. In contrast, the decrease in the number dying during the first year of life out of 1000 new born appears to have picked up the momentum after 2001–2005.

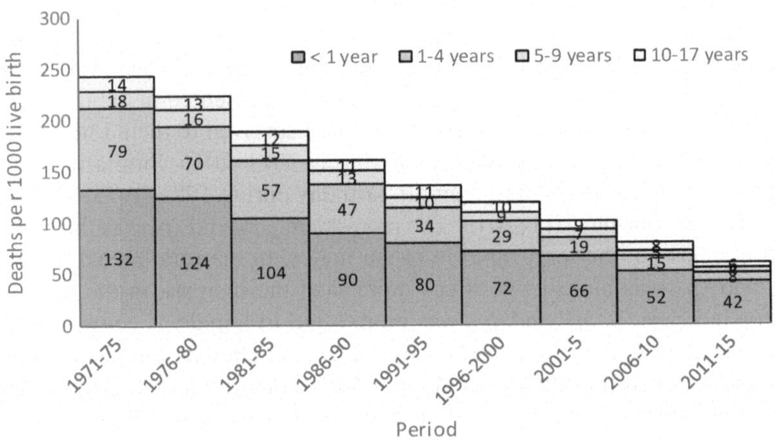

Fig. 5.1 Distribution of child deaths (per 1000 live births) in India

Table 5.2 Residence and gender effects of the improvement in child survival probability in India

Effects		Survival probability					
		$_{1}p_{0}$	$_{4}p_{1}$	$_{5}p_{0}$	$_{5}p_{5}$	$_{8}p_{10}$	$_{18}p_{0}$
Average improvement		0.076	0.064	0.131	0.016	0.010	0.150
Residence effects	Rural	0.019	0.027	0.042	0.006	0.004	0.047
	Urban	−0.019	−0.027	−0.042	−0.006	−0.004	−0.047
Gender effects	Male	0	−0.012	−0.010	−0.002	−0.002	−0.012
	Female	0	0.012	0.010	0.002	0.002	−0.012
Residuals	Rural male	−0.002	0	−0.002	−0.001	−0.001	−0.003
	Rural female	0.002	0	0.002	0.001	0.001	0.003
	Urban male	0.002	0	0.002	0.001	0.001	0.003
	Urban female	−0.002	0	−0.002	−0.001	−0.001	−0.003

Source Author's calculations

We have applied the mean polish technique (Selvin 2004) to estimate the gender and residence effects of the improvement in child survival probability at different ages of the childhood and the results are presented in Table 5.2. The table suggests that improvement in rural $_{18}p_{0}$ is higher by 0.047 absolute points than the average $_{18}p_{0}$, which means that improvement in urban $_{18}p_{0}$ is lower by 0.047 absolute points than the average. Similarly, improvement in female $_{18}p_{0}$ is higher by 0.012 absolute points than the average, which means that improvement in male $_{18}p_{0}$ is lower by 0.012 absolute points than the average. The table also reveals that improvement in rural $_{5}p_{0}$ is higher by 0.042 absolute points or improvement in urban $_{5}p_{0}$ is lower by 0.042 absolute points than the average, whereas improvement in female $_{5}p_{0}$ is higher by 0.010 absolute points or improvement in male $_{5}p_{0}$ is lower by 0.010 absolute points than the average.

It may also be seen from Table 5.2 that the residence and gender effects of the improvement in $_{1}p_{0}$ are quite small compared to residence and gender effects of the improvement in $_{4}p_{1}$. On the other hand, residence effects of the improvement in $_{5}p_{5}$ are larger than that in $_{8}p_{10}$ but gender effects of the improvement in $_{8}p_{10}$ are larger than that in $_{5}p_{5}$. In any case, it is obvious from Table 5.2 that the difference in improvement in survival probability across the four mutually exclusive population sub-groups increases with the age of the child. They are quite narrow in $_{1}p_{0}$ but quite substantial in $_{18}p_{0}$. More importantly, the improvement in female $_{1}p_{0}$ has been slower than that in male $_{1}p_{0}$ but not in $_{4}p_{1}$, $_{5}p_{5}$ and $_{8}p_{10}$.

The pace of improvement in the child survival probability has, however, not been the same throughout the 40 years under reference and there has been considerable slowdown in the improvement in the post 1991–1995 period compared to the period before 1991–1995. During 1971–1975 through 1991–1995, $_{18}p_{0}$ improved by 0.109 absolute points whereas the improvement during 1991–1995 through 2011–2015 was only around 0.075 absolute points. The slowdown in the improvement is particularly marked in $_{4}p_{1}$. The improvement in $_{1}p_{0}$ gained some momentum after 2001–2005 but the improvement in survival probability at other ages of the childhood continued to slowdown after 1991–1995.

The slowdown in the improvement in child survival probability during the post 1991–1995 period appears to be the reason behind the country missing the MDG-4 goal of reducing the under-five mortality rate by two-thirds between 1990 and 2015 as confirmed by the official report on the progress towards MDGs prepared by the Government of India (Government of India 2015a). The under-five mortality rate in India in 1990 was estimated to be around 126 under-five deaths per 1000 live births, which means that ${}_5p_0$ should have been improved to 0.998 by the year 2015 to achieve the MDG-4. If the improvement in ${}_5p_0$ during 1991–1995 through 2011–2015 would have been the same as the improvement during 1971–1975 through 1991–1995, then ${}_5p_0$ in the country would have increased to 0.983 during 2011–2015 in which case that the country would have been able to achieve the MDG-4 by the year 2015. However, instead of accelerating, the improvement in the probability of survival during the first 5 years, ${}_5p_0$, decelerated during the post 1991–1995 period so that ${}_5p_0$ could increase to only 0.949 during the period 2011–2015. The primary reason behind the slowdown in the increase in ${}_5p_0$ has been the slowdown in the improvement in the probability of survival during 1–4 years of life, ${}_4p_1$.

5.5 Decomposition of the Improvement in Child Survival Probability

Table 5.3 presents the relative contribution of the increase in ${}_1p_0$, ${}_4p_1$, ${}_5p_5$ and ${}_8p_{10}$ to the increase in ${}_{18}p_0$ in conjunction with the Eq. (5.5) as well as the contribution of the increase in ${}_1p_0$ and ${}_4p_1$, to the increase in ${}_5p_0$ in conjunction with the Eq. (5.6). It may be seen from the table that around 45% of the improvement in ${}_{18}p_0$ in the country between 1971–1975 and 2011–2015 was the result of the improvement in ${}_1p_0$ while another 40% was the result of the improvement in ${}_4p_1$ so that around 85% of the improvement in ${}_{18}p_0$ was the result of the improvement in the probability of survival during the first 5 years of life, ${}_5p_0$, during the 40 years under reference. The remaining 15% improvement was the result of the improvement in ${}_5p_5$ and ${}_8p_{10}$. On the other hand, 53% of the improvement in ${}_5p_0$ was the result of the improvement in ${}_1p_0$ and around 47% to the improvement in ${}_4p_1$.

The relative contribution of the improvement in ${}_1p_0$ and ${}_4p_1$ to the improvement in ${}_{18}p_0$ and ${}_5p_0$, however, varies by residence and gender. In rural male children, 47% of the improvement in ${}_{18}p_0$ and 55% of the improvement in ${}_5p_0$ are attributed to the improvement in ${}_1p_0$, whereas these proportions are 42% and 50%, respectively, in rural female children. In urban male children, 61% of the improvement in ${}_{18}p_0$ and 71% of the improvement in ${}_5p_0$ are the results of the improvement in ${}_1p_0$, but these proportions are only 46% and 54%, respectively, in urban female children. On the

Table 5.3 Decomposition of the improvement in ${}_{18}p_0$ and ${}_{5}p_0$ in India, 1971–2015

Period	Improvement in	Proportionate (per cent) improvement attributed to improvement in				Improvement in	Proportionate improvement attributed to improvement in	
	${}_{18}p_0$	${}_{1}p_0$	${}_{4}p_1$	${}_{5}p_5$	${}_{8}p_{10}$	${}_{5}p_0$	${}_{1}p_0$	${}_{4}p_1$
Combined population, both sexes								
1971–1975/1976–1980	0.02	34.26	45.27	12.62	7.85	0.02	43.08	56.92
1976–1980/1981–1985	0.03	52.06	38.91	5.61	3.43	0.03	57.23	42.77
1981–1985/1986–1990	0.03	46.59	39.06	9.20	5.16	0.02	54.40	45.60
1986–1990/1991–1995	0.03	35.44	52.11	8.75	3.69	0.02	40.48	59.52
1991–1995/1996–2000	0.02	50.96	30.00	11.28	7.76	0.01	62.94	37.06
1996–2000/2001–2005	0.02	30.30	53.58	10.67	5.45	0.02	36.12	63.88
2001–2005/2006–2010	0.02	65.75	19.27	7.77	7.21	0.02	77.33	22.67
2006–2010/2011–2015	0.02	49.30	33.19	8.62	8.89	0.02	59.76	40.24
1971–1975/2011–2015	*0.18*	*45.39*	*39.85*	*8.98*	*5.77*	*0.16*	*53.25*	*46.75*
Combined population, male								
1971–1975/1976–1980	0.02	37.87	44.39	11.90	5.84	0.02	46.04	53.96
1976–1980/1981–1985	0.03	51.31	37.78	6.64	4.27	0.03	57.60	42.40
1981–1985/1986–1990	0.03	54.61	33.05	8.48	3.87	0.03	62.29	37.71
1986–1990/1991–1995	0.02	29.20	55.39	10.65	4.76	0.02	34.52	65.48
1991–1995/1996–2000	0.02	54.20	29.02	10.11	6.66	0.01	65.13	34.87
1996–2000/2001–2005	0.02	42.82	45.44	8.17	3.58	0.01	48.52	51.48
2001–2005/2006–2010	0.02	65.91	19.21	7.74	7.15	0.02	77.44	22.56
2006–2010/2011–2015	0.02	54.03	30.31	8.52	7.14	0.02	64.06	35.94
1971–1975/2011–2015	*0.17*	*48.77*	*37.12*	*8.89*	*5.22*	*0.15*	*56.78*	*43.22*

(continued)

Table 5.3 (continued)

Period	Improvement in	Proportionate (per cent) improvement attributed to improvement in				Improvement in	Proportionate improvement attributed to improvement in	
	18P0	*1P0*	*4P1*	*5P5*	*8P10*	*5P0*	*1P0*	*4P1*
Combined population, female								
1971–1975/1976–1980	0.02	28.87	50.16	10.72	10.26	0.02	36.53	63.47
1976–1980/1981–1985	0.04	53.94	37.65	5.75	2.65	0.04	58.89	41.11
1981–1985/1986–1990	0.03	50.17	36.26	8.98	4.59	0.03	58.05	41.95
1986–1990/1991–1995	0.03	24.82	62.59	8.65	3.95	0.02	28.39	71.61
1991–1995/1996–2000	0.01	50.30	26.43	12.89	10.38	0.01	65.56	34.44
1996–2000/2001–2005	0.02	19.83	61.42	12.22	6.53	0.02	24.41	75.59
2001–2005/2006–2010	0.02	65.60	18.94	8.18	7.28	0.02	77.60	22.40
2006–2010/2011–2015	0.02	45.13	35.98	8.43	10.46	0.02	55.65	44.35
1971–1975/2011–2015	*0.20*	*42.29*	*42.43*	*9.00*	*6.28*	*0.17*	*49.92*	*50.08*
Rural population, both sexes								
1971–1975/1976–1980	0.02	27.45	51.43	12.79	8.33	0.02	34.80	65.20
1976–1980/1981–1985	0.03	55.38	36.37	4.88	3.37	0.03	60.36	39.64
1981–1985/1986–1990	0.03	41.69	42.74	10.48	5.09	0.02	49.38	50.62
1986–1990/1991–1995	0.03	39.85	48.87	7.93	3.35	0.03	44.92	55.08
1991–1995/1996–2000	0.02	43.29	38.87	11.42	6.42	0.02	52.69	47.31
1996–2000/2001–2005	0.02	29.76	51.11	11.80	7.32	0.01	36.80	63.20
2001–2005/2006–2010	0.02	67.05	18.39	7.49	7.07	0.02	78.48	21.52
2006–2010/2011–2015	0.02	46.88	33.90	9.42	9.80	0.02	58.03	41.97
1971–1975/2011–2015	*0.20*	*44.56*	*40.54*	*9.05*	*5.86*	*0.17*	*52.36*	*47.64*

(continued)

Table 5.3 (continued)

Period	Improvement in	Proportionate (per cent) improvement attributed to improvement in				Improvement in	Proportionate improvement attributed to improvement in	
	$_{18}p_0$	$_1p_0$	$_4p_1$	$_5p_5$	$_8p_{10}$	$_5p_0$	$_1p_0$	$_4p_1$
Rural population, male								
1971–1975/1976–1980	0.02	32.79	50.40	11.09	5.72	0.02	39.42	60.58
1976–1980/1981–1985	0.03	53.27	36.03	6.52	4.17	0.03	59.65	40.35
1981–1985/1986–1990	0.03	43.09	42.31	10.01	4.59	0.02	50.46	49.54
1986–1990/1991–1995	0.03	44.03	44.10	8.78	3.10	0.03	49.96	50.04
1991–1995/1996–2000	0.02	49.20	34.24	10.28	6.28	0.02	58.97	41.03
1996–2000/2001–2005	0.01	41.67	44.50	8.98	4.85	0.01	48.36	51.64
2001–2005/2006–2010	0.02	66.82	19.41	7.30	6.47	0.02	77.49	22.51
2006–2010/2011–2015	0.02	50.67	31.78	9.48	8.06	0.02	61.45	38.55
1971–1975/2011–2015	*0.18*	*47.49*	*38.37*	*8.95*	*5.18*	*0.16*	*55.31*	*44.69*
Rural population, female								
1971–1975/1976–1980	0.02	26.77	50.34	10.57	12.32	0.02	34.71	65.29
1976–1980/1981–1985	0.04	55.68	37.40	5.07	1.86	0.04	59.82	40.18
1981–1985/1986–1990	0.03	40.49	42.17	11.40	5.94	0.02	48.98	51.02
1986–1990/1991–1995	0.04	36.26	54.07	6.82	2.85	0.03	40.14	59.86
1991–1995/1996–2000	0.02	37.86	40.71	12.98	8.45	0.01	48.19	51.81
1996–2000/2001–2005	0.02	20.51	57.48	13.36	8.65	0.02	26.30	73.70
2001–2005/2006–2010	0.02	67.19	17.64	8.13	7.04	0.02	79.20	20.80
2006–2010/2011–2015	0.02	43.49	35.77	9.18	11.56	0.02	54.87	45.13
1971–1975/2011–2015	*0.21*	*41.82*	*42.58*	*9.07*	*6.53*	*0.18*	*49.54*	*50.46*

(continued)

Table 5.3 (continued)

Period	Improvement in	Proportionate (per cent) improvement attributed to improvement in				Improvement in	Proportionate improvement attributed to improvement in	
	18P0	*1P0*	*4P1*	*5P5*	*8P10*	*5P0*	*1P0*	*4P1*
Urban population, both sexes								
1971–1975/1976–1980	0.01	92.24	−16.75	13.99	10.52	0.01	122.19	−22.19
1976–1980/1981–1985	0.02	48.62	44.06	6.90	0.43	0.02	52.46	47.54
1981–1985/1986–1990	0.01	40.68	44.17	6.17	8.99	0.01	47.94	52.06
1986–1990/1991–1995	0.01	38.02	48.30	9.91	3.78	0.01	44.04	55.96
1991–1995/1996–2000	0.01	73.64	9.16	6.34	10.86	0.01	88.94	11.06
1996–2000/2001–2005	0.01	30.73	58.05	10.17	1.06	0.01	34.61	65.39
2001–2005/2006–2010	0.01	60.29	20.83	8.14	10.73	0.01	74.32	25.68
2006–2010/2011–2015	0.01	59.68	32.57	3.99	3.75	0.01	64.69	35.31
1971–1975/2011–2015	*0.11*	*55.09*	*30.89*	*8.31*	*5.70*	*0.09*	*64.07*	*35.93*
Urban population, male								
1971–1975/1976–1980	0.01	140.99	−99.32	29.05	29.28	0.00	338.36	−238.40
1976–1980/1981–1985	0.01	72.33	28.71	3.94	−4.98	0.01	71.59	28.41
1981–1985/1986–1990	0.01	13.20	64.47	9.71	12.62	0.01	16.99	83.01
1986–1990/1991–1995	0.01	53.62	38.14	4.98	3.27	0.01	58.43	41.57
1991–1995/1996–2000	0.01	62.92	19.84	9.36	7.88	0.01	76.02	23.98
1996–2000/2001–2005	0.01	51.54	41.89	6.20	0.37	0.01	55.16	44.84
2001–2005/2006–2010	0.01	63.28	15.96	10.92	9.84	0.01	79.85	20.15
2006–2010/2011–2015	0.01	69.99	25.14	3.11	1.76	0.01	73.57	26.43
1971–1975/2011–2015	*0.09*	*60.51*	*24.56*	*8.60*	*6.32*	*0.08*	*71.13*	*28.87*

(continued)

Table 5.3 (continued)

Period	Improvement in	Proportionate (per cent) improvement attributed to improvement in				Improvement in	Proportionate improvement attributed to improvement in	
	18P0	*1P0*	*4P1*	*5P5*	*8P10*	*5P0*	*1P0*	*4P1*
Urban population, female								
1971–1975/1976–1980	0.01	55.60	21.06	15.07	8.27	0.01	72.53	27.47
1976–1980/1981–1985	0.02	51.42	38.80	7.60	2.18	0.02	57.00	43.00
1981–1985/1986–1990	0.02	45.21	45.00	4.71	5.08	0.02	50.12	49.88
1986–1990/1991–1995	0.01	20.82	59.68	14.10	5.41	0.01	25.86	74.14
1991–1995/1996–2000	0.01	76.94	−0.03	9.08	14.01	0.01	100.03	−0.03
1996–2000/2001–2005	0.01	14.17	74.78	10.12	0.93	0.01	15.93	84.07
2001–2005/2006–2010	0.01	58.18	24.09	7.58	10.16	0.01	70.72	29.28
2006–2010/2011–2015	0.01	49.62	40.35	3.81	6.23	0.01	55.15	44.85
1971–1975/2011–2015	*0.11*	*46.02*	*39.49*	*8.75*	*5.74*	*0.10*	*53.82*	*46.18*

Source Author's calculations

other hand, the contribution of the improvement in ${}_4p_1$ to the improvement in ${}_{18}p_0$ and ${}_5p_0$ is relatively larger in the rural than in the urban areas of the country.

It is also evident from Table 5.3 that the contribution of the improvement in ${}_4p_1$ to the improvement in ${}_{18}p_0$ and ${}_5p_0$ is substantially smaller during the post 1991–1995 period, as compared to the pre 1991–1995 period. This observation again indicates that improvement in ${}_4p_1$ slowed down during the post 1991–1995 period. This slowdown appears to be primarily the reason behind the slowdown in the improvement in both ${}_{18}p_0$ and ${}_5p_0$ during the post 1991–1995 period, as compared to the pre 1991–1995 period. The slowdown in ${}_4p_1$ also appears to be reason behind India missing the MDG-4.

5.6 Inter-state Variations

Any discussion on child survival progress in India is incomplete without an analysis of within country, inter-state, variation in the progress towards child survival. Estimates of childhood survival probabilities during 2011–2015 for 17 states of the country derived from the official life tables are presented in Table 5.4, which shows wide

Table 5.4 Child survival probabilities in Indian states, 2011–2015

State	${}_1p_0$	${}_4p_1$	${}_5p_0$	${}_5p_5$	${}_8p_{10}$	${}_{18}p_0$
Andhra Pradesh	0.932	0.998	0.930	0.998	0.995	0.923
Assam	0.950	0.981	0.932	0.995	0.992	0.920
Bihar	0.963	0.988	0.951	0.995	0.994	0.940
Gujarat	0.956	0.991	0.947	0.996	0.994	0.939
Haryana	0.956	0.994	0.950	0.997	0.995	0.943
Himachal Pradesh	0.960	0.995	0.955	0.998	0.996	0.949
Jammu & Kashmir	0.947	0.997	0.944	0.997	0.995	0.936
Karnataka	0.968	0.996	0.964	0.997	0.994	0.956
Kerala	0.988	0.999	0.987	0.999	0.997	0.984
Madhya Pradesh	0.941	0.984	0.926	0.994	0.991	0.912
Maharashtra	0.980	0.997	0.977	0.998	0.995	0.970
Odisha	0.953	0.986	0.940	0.994	0.991	0.925
Punjab	0.974	0.995	0.969	0.998	0.994	0.962
Rajasthan	0.945	0.990	0.936	0.996	0.994	0.926
Tamil Nadu	0.979	0.998	0.977	0.998	0.994	0.969
Uttar Pradesh	0.937	0.987	0.925	0.995	0.993	0.914
West Bengal	0.971	0.996	0.967	0.997	0.994	0.959

Source Author's calculations

inter-state variation in child survival as ${}_{18}p_0$ ranges from the highest in Kerala to the lowest in Madhya Pradesh. Kerala also has the highest survival probability in all ages of the childhood period, whereas ${}_{1}p_0$ is the lowest in Uttar Pradesh and ${}_{5}p_{10}$ is the lowest in Odisha. At other ages of the childhood period, the survival probability is the lowest in Madhya Pradesh.

The improvement in ${}_{18}p_0$ and ${}_{5}p_0$ between 1971–1975 and 2011–2015 has also been different in different states (Table 5.5). The improvement in ${}_{18}p_0$ has been most rapid in Uttar Pradesh (40%) and quite rapid in Madhya Pradesh and Rajasthan but least rapid in Karnataka (8%) and slow in Haryana, Jammu and Kashmir and Punjab. The improvement in ${}_{1}p_0$ and ${}_{4}p_1$ has also been the most rapid in Uttar Pradesh but improvement in ${}_{5}p_5$ and ${}_{8}p_{10}$ has been the most rapid in Assam. On the other hand, improvement in survival probability has been the slowest in Karnataka and slow in Haryana.

The contribution of the improvement in ${}_{1}p_0$, ${}_{4}p_1$, ${}_{5}p_5$ and ${}_{8}p_{10}$ to the improvement in ${}_{18}p_0$ also varies across the states of the country for which life tables have been prepared by the Registrar General and Census Commissioner of India on the basis of the Sample Registration System. In Punjab and Gujarat, for example, around 90% of the improvement in ${}_{18}p_0$ during the period under reference is attributed to the improve-

Table 5.5 Decomposition of the improvement in child survival probability in selected states of India, 1971–2015

State	Improvement in	Proportionate improvement in ${}_{18}p_0$ attributed to the improvement in				Improvement in	Proportionate improvement in ${}_{5}p_0$ attributed to the improvement in	
	${}_{18}p_0$	${}_{1}p_0$	${}_{4}p_1$	${}_{5}p_5$	${}_{8}p_{10}$	${}_{5}p_0$	${}_{1}p_0$	${}_{4}p_1$
Andhra Pradesh	0.156	25.8	53.1	12.6	8.5	0.126	32.7	67.3
Assam	0.180	44.2	28.6	14.1	13.1	0.137	60.7	39.3
Bihar	na	na	na	na	na	na	na	na
Gujarat	0.199	45.1	44.6	6.2	4.1	0.182	50.3	49.7
Haryana	0.125	37.7	47.4	9.0	5.9	0.108	44.3	55.7
Himachal Pradesh	na	na	na	na	na	na	na	na
Jammu & Kashmir	0.120	28.3	51.0	11.8	8.9	0.097	35.7	64.3
Karnataka	0.071	33.0	53.3	10.6	3.1	0.062	38.3	61.7
Kerala	na	na	na	na	na	na	na	na
Madhya Pradesh	0.197	39.7	46.8	7.8	5.7	0.175	45.9	54.1
Maharashtra	0.168	44.3	42.1	8.4	5.2	0.148	51.3	48.7
Odisha	0.192	43.7	38.5	9.9	7.9	0.164	53.1	46.9
Punjab	0.141	54.5	37.1	5.3	3.1	0.131	59.5	40.5
Rajasthan	0.199	38.2	48.3	9.1	4.4	0.177	44.2	55.8
Tamil Nadu	0.197	43.8	41.4	8.6	6.2	0.172	51.4	48.6
Uttar Pradesh	0.259	38.7	48.5	8.7	4.1	0.233	44.4	55.6
West Bengal	na	na	na	na	na	na	na	na

Source Author's calculations

ment in ${}_1p_0$ and ${}_4p_1$, whereas in Assam, this proportion is found to be less than 75%. In Punjab, almost 55% of the improvement in ${}_{18}p_0$ is attributed to the improvement in ${}_1p_0$ alone. On the other hand, in seven states of the country, this proportion is less than 40%. In Jammu and Kashmir and Karnataka, improvement in ${}_1p_0$ accounts for only about one-third of the improvement in ${}_{18}p_0$. Similarly, improvement in ${}_4p_1$ accounts for more than 53% of the improvement in ${}_{18}p_0$ in Karnataka but only 28% in Assam during the period under reference. In majority of the states of the country, however, about 85–87% of the improvement in the child survival probability (${}_{18}p_0$) during the period under reference is accounted by the improvement in the survival probability in the first 5 years of life (${}_1p_0$ and ${}_4p_1$). Assam is the only state in the country where more than 25% of the improvement in ${}_{18}p_0$ is attributed to the improvement in survival probability in children at least 5 years of age. In any case, it is obvious from Table 5.4 that the pathway of improvement in the childhood survival probability in different states of the country during the period under reference has essentially been different.

As regards MDG-4, there are only three states—Gujarat, Maharashtra and Tamil Nadu—which have been able to reduce the under-five mortality rate by at least two-thirds between 1990 and 2013 (2011–2015). Moreover, Karnataka and West Bengal are likely to have achieved MDG-4 within one or two years of 2015 (Table 5.6). On the other hand, distance to MDG-4 appears to be the farthest in Uttar Pradesh where ${}_5p_0$ is estimated to be 0.924 against the 2015 goal of 0.945. Other states which may have missed MDG-4 are Haryana, Himachal Pradesh, Madhya Pradesh and Rajasthan. In many of these states, improvement in ${}_5p_0$ has been quite rapid but the pace of improvement appears to be insufficient to achieve the MDG-4. In the context of the National Health Policy 2017 and the United Nations Sustainable Development Goals, it is obvious that these states will have to make significant additional efforts to improve ${}_5p_0$ to more than 0.975 by the year 2030.

5.7 Discussion and Conclusions

The foregoing analysis reflects a rather unsatisfactory progress towards securing the life of Indian children during the 40 years under reference. A disturbing finding of the present analysis is that the child survival progress slowed down after 1991–1995, which appears to be the reason behind India missing the MDG-4. Had the pace of improvement in the child survival probability during 1971–1975 through 1991–1995 been maintained during 1991–1995 through 2011–2015, India would have been able to achieve the MDG-4. Although, there has been some acceleration in the improvement in child survival probability after 2001–2005, yet this acceleration is at best marginal and is confined to the first year of life only. The slowdown in the improvement has been particularly marked in the 1–4 years of life. The recent report by UNICEF also confirms that progress towards child survival in India has been slow by international standards (UNICEF 2016).

Table 5.6 Progress towards MDG-4 in India and states

Country/state	${}_5p_0$ in 1990	Under-five mortality rate in 1990	MDG-4 target	${}_5p_0$ in 2011–2015	Under-five mortality rate in 2011–2015
India	0.874	0.126	0.042	0.950	0.050
Andhra Pradesh	0.904	0.096	0.032	0.961	0.039
Assam	0.854	0.146	0.049	0.932	0.068
Bihar	0.660	0.340	0.113	0.951	0.049
Gujarat	0.803	0.197	0.066	0.948	0.052
Haryana	0.895	0.105	0.035	0.951	0.049
Himachal Pradesh	0.910	0.090	0.030	0.955	0.045
Jammu & Kashmir	0.896	0.104	0.035	0.955	0.045
Karnataka	0.896	0.104	0.035	0.964	0.036
Kerala	0.973	0.027	0.009	0.987	0.013
Madhya Pradesh	0.818	0.182	0.061	0.926	0.074
Maharashtra	0.919	0.081	0.027	0.977	0.023
Odisha	0.840	0.160	0.053	0.940	0.060
Punjab	0.916	0.084	0.028	0.969	0.031
Rajasthan	0.849	0.151	0.050	0.936	0.064
Tamil Nadu	0.919	0.081	0.027	0.977	0.023
Uttar Pradesh	0.835	0.165	0.055	0.924	0.076
West Bengal	0.902	0.098	0.033	0.966	0.034

Source Author's calculations

The present analysis also suggests that the gender and residence differentials in child survival probability in the country have narrowed down over time. However, this narrowing of gender and residence differentials appears to be the result of relatively slow improvement in the survival probability of male as compared to female children in the urban as compared to the rural areas so that the improvement in the survival probability has been the slowest in urban male children but the fastest in the rural female children. In the urban areas of the country, the probability of survival in the first 5 years of life and in the first 18 years of life is now higher in female than in male children because of the slow improvement in the survival probability of male children. Even in the rural areas, improvement in the survival probability of female children has been faster than that in male children so that the gender difference in child survival probability has narrowed down considerably, although male children survival probability remains higher than female children survival probability.

Reasons for differential improvement in the child survival probability in India are not known at present. It is generally argued that health care services delivery infrastructure is better in urban than in rural areas. Similarly, the general level of social and economic development is argued to be better in urban than in rural areas. The evidence available from the latest round of the National Family Health Survey 2015–2016 also indicates that the coverage of key child survival interventions, such as immunisation against vaccine preventable diseases, use of oral rehydration therapy during diarrhoea, initiation of breastfeeding within 1 hour of birth, and so on, is relatively higher in urban than in rural areas of the country. Similarly, the prevalence of child under nutrition measured in terms of the prevalence of stunting, wasting and under-weight has also been found to be higher in rural than in urban areas currently as well as in the past. It is also argued that females face different forms of discrimination at the level of the family and the society which has implications for the survival of female children. However, despite numerous urban and many male advantages, the slow improvement in child survival probability in male children, particularly in urban male children is a matter of concern that needs comprehensive scrutiny.

The history of child survival efforts in India can be divided into three distinct phases. The first phase started with the independence and lasted up to 1977. The child survival and health-related services, during this phase, were delivered through the public health institutions, although a community-based programme of training of traditional birth attendants was launched in 1955 to prevent child deaths from neonatal tetanus. The impact of these services on child survival, however, appears to be limited, at best, as is reflected from the very low child survival probabilities during 1971–1975.

The second phase of child survival efforts spanned from 1975 through 1992. This phase focused on specific child survival interventions such as immunisation and oral rehydration therapy, including community mobilisation for the cause of child survival. The most notable programmatic intervention of this phase was the Universal Immunisation Programme which was launched in 1985 and received the highest priority of the government by according the status of National Technology Mission in 1987. The Integrated Child Development Scheme was also launched during this period that focused on early childhood development.

The third phase of child survival efforts had a beginning in 1992. This phase focused on an integrated approach in which child survival activities were integrated with reproductive health and other services in line with the concept of continuum of care. However, improvement in child survival probability during this phase has been considerably slower than the improvement in the earlier phase. Integration of child survival activities with reproductive and other health care services appears to have resulted in a residual attention paid to key child survival interventions so that improvement in child survival probability have slowed down during this phase.

From the global perspective, an accelerated improvement in child survival probability in India is the need of the time. At the current rate of improvement, total number of under-five deaths in India will be accounting for 17% of the global under-five deaths in 2030. Obviously, the pace of reduction in the under-five mortality rate in India will influence substantially the global progress towards realising the target of reducing the under-five mortality rate at least as low as 25 under-five deaths per 1000 live births by 2030 as articulated in the 2030 Sustainable Development Agenda (United Nations 2015). The under-five mortality rate in India needs to decrease by 3–4% per year in the coming years to achieve the target of reducing under-five mortality rate to 23 under-five deaths per 1000 live births by 2025, infant mortality rate to 28 infant deaths per 1000 live births by 2019 and neonatal mortality rate to 16 neonatal deaths per 1000 live births by 2025 as laid down in the National Health Policy 2017. The task appears to be challenging as it is projected that, at the current pace of reduction, the under-five mortality rate will reduce to only 31 under-five deaths per 1000 live births, while the neonatal mortality rate will decrease to only 20 neonatal deaths per 1000 live births by 2025. On the other hand, it is projected that the infant mortality rate will decrease to only 33 infant deaths per 1000 live births by 2019 (De et al. 2016).

Appendix

See Appendix Table A.5.1.

Table A.5.1 Child survival probabilities in India, 1971–2015

Period	Survival probability					
	$_1p_0$	$_4p_1$	$_5p_0$	$_5p_5$	$_8p_{10}$	$_{18}p_0$
Combined population, both sexes						
1971–1975	0.868	0.909	0.789	0.977	0.981	0.756
1976–1980	0.876	0.920	0.806	0.980	0.983	0.777
1981–1985	0.896	0.936	0.839	0.983	0.985	0.812
1986–1990	0.910	0.948	0.863	0.986	0.987	0.839
1991–1995	0.920	0.963	0.886	0.988	0.988	0.865
1996–2000	0.928	0.969	0.899	0.990	0.989	0.881
2001–2005	0.935	0.980	0.916	0.992	0.990	0.900
2006–2010	0.948	0.984	0.933	0.994	0.992	0.919
2011–2015	0.958	0.991	0.949	0.996	0.994	0.940
Improvement (%)	*10.369*	*9.021*	*20.320*	*1.945*	*1.325*	*24.339*

(continued)

Table A.5.1 (continued)

Period	Survival probability					
	$_1p_0$	$_4p_1$	$_5p_0$	$_5p_5$	$_8p_{10}$	$_{18}p_0$
Combined population, male						
1971–1975	0.871	0.922	0.803	0.979	0.983	0.773
1976–1980	0.879	0.933	0.820	0.982	0.985	0.792
1981–1985	0.896	0.946	0.848	0.984	0.986	0.823
1986–1990	0.913	0.957	0.873	0.987	0.988	0.851
1991–1995	0.919	0.970	0.891	0.989	0.989	0.872
1996–2000	0.929	0.975	0.905	0.991	0.990	0.888
2001–2005	0.936	0.983	0.920	0.993	0.991	0.904
2006–2010	0.948	0.987	0.936	0.994	0.992	0.923
2011–2015	0.959	0.993	0.952	0.996	0.994	0.942
Improvement (%)	*10.163*	*7.642*	*18.580*	*1.778*	*1.119*	*21.946*
Combined population, female						
1971–1975	0.866	0.895	0.774	0.975	0.979	0.739
1976–1980	0.873	0.908	0.793	0.978	0.982	0.762
1981–1985	0.896	0.925	0.829	0.981	0.984	0.800
1986–1990	0.913	0.938	0.857	0.984	0.985	0.831
1991–1995	0.921	0.957	0.881	0.987	0.987	0.858
1996–2000	0.928	0.961	0.892	0.989	0.988	0.871
2001–2005	0.933	0.977	0.911	0.992	0.990	0.895
2006–2010	0.947	0.981	0.929	0.994	0.991	0.915
2011–2015	0.957	0.989	0.946	0.996	0.994	0.937
Improvement (%)	*10.534*	*10.571*	*22.220*	*2.152*	*1.532*	*26.714*
Rural population, both sexes						
1971–1975	0.858	0.899	0.772	0.975	0.980	0.737
1976–1980	0.865	0.912	0.789	0.978	0.982	0.758
1981–1985	0.887	0.928	0.823	0.980	0.984	0.793
1986–1990	0.900	0.941	0.847	0.984	0.985	0.821
1991–1995	0.914	0.959	0.876	0.987	0.987	0.853
1996–2000	0.922	0.967	0.892	0.989	0.988	0.871
2001–2005	0.928	0.977	0.907	0.992	0.989	0.889
2006–2010	0.943	0.981	0.925	0.993	0.991	0.911
2011–2015	0.953	0.989	0.943	0.996	0.993	0.933
Improvement (%)	*11.077*	*10.029*	*22.220*	*2.152*	*1.327*	*26.581*

(continued)

Table A.5.1 (continued)

Period	Survival probability					
	$_{1}p_{0}$	$_{4}p_{1}$	$_{5}p_{0}$	$_{5}p_{5}$	$_{8}p_{10}$	$_{18}p_{0}$
Rural population, male						
1971–1975	0.861	0.913	0.786	0.977	0.982	0.754
1976–1980	0.869	0.926	0.805	0.980	0.984	0.776
1981–1985	0.887	0.939	0.833	0.982	0.985	0.806
1986–1990	0.899	0.952	0.856	0.985	0.987	0.832
1991–1995	0.913	0.966	0.882	0.988	0.988	0.861
1996–2000	0.923	0.973	0.898	0.990	0.989	0.879
2001–2005	0.929	0.980	0.911	0.992	0.990	0.894
2006–2010	0.943	0.985	0.929	0.993	0.991	0.915
2011–2015	0.954	0.992	0.946	0.996	0.993	0.936
Improvement (%)	*10.788*	*8.626*	*20.350*	*1.947*	*1.120*	*24.065*
Rural population, female						
1971–1975	0.855	0.884	0.756	0.973	0.977	0.719
1976–1980	0.862	0.898	0.775	0.976	0.981	0.742
1981–1985	0.887	0.916	0.812	0.978	0.982	0.780
1986–1990	0.900	0.929	0.836	0.982	0.984	0.808
1991–1995	0.914	0.952	0.870	0.985	0.985	0.844
1996–2000	0.922	0.960	0.884	0.988	0.987	0.862
2001–2005	0.926	0.974	0.902	0.991	0.989	0.884
2006–2010	0.942	0.978	0.921	0.993	0.991	0.906
2011–2015	0.952	0.987	0.9400	0.995	0.994	0.930
Improvement (%)	*11.373*	*11.594*	*24.290*	*2.359*	*1.740*	*29.370*
Urban population, both sexes						
1971–1975	0.911	0.961	0.875	0.987	0.988	0.854
1976–1980	0.926	0.958	0.887	0.990	0.990	0.869
1981–1985	0.936	0.968	0.906	0.991	0.990	0.889
1986–1990	0.942	0.974	0.918	0.992	0.991	0.903
1991–1995	0.947	0.980	0.928	0.994	0.992	0.915
1996–2000	0.955	0.981	0.937	0.994	0.993	0.925
2001–2005	0.959	0.990	0.950	0.996	0.993	0.939
2006–2010	0.965	0.992	0.958	0.997	0.994	0.949
2011–2015	0.973	0.996	0.969	0.997	0.995	0.961
Improvement (%)	*6.747*	*3.727*	*10.730*	*0.989*	*0.709*	*12.580*

(continued)

Table A.5.1 (continued)

Period	Survival probability					
	${}_{1}p_{0}$	${}_{4}p_{1}$	${}_{5}p_{0}$	${}_{5}p_{5}$	${}_{8}p_{10}$	${}_{18}p_{0}$
Urban population, male						
1971–1975	0.915	0.971	0.889	0.989	0.988	0.869
1976–1980	0.926	0.964	0.892	0.991	0.991	0.876
1981–1985	0.936	0.968	0.906	0.991	0.990	0.889
1986–1990	0.938	0.978	0.918	0.993	0.992	0.904
1991–1995	0.945	0.983	0.929	0.994	0.993	0.917
1996–2000	0.954	0.986	0.940	0.995	0.994	0.930
2001–2005	0.960	0.992	0.952	0.996	0.994	0.943
2006–2010	0.966	0.993	0.960	0.997	0.995	0.951
2011–2015	0.974	0.996	0.970	0.997	0.995	0.963
Improvement (%)	*6.420*	*2.557*	*9.140*	*0.888*	*0.709*	*10.828*
Urban population, female						
1971–1975	0.917	0.949	0.870	0.987	0.988	0.848
1976–1980	0.925	0.952	0.881	0.989	0.989	0.861
1981–1985	0.938	0.962	0.902	0.991	0.990	0.885
1986–1990	0.946	0.970	0.917	0.992	0.990	0.901
1991–1995	0.948	0.977	0.927	0.993	0.991	0.912
1996–2000	0.956	0.977	0.934	0.994	0.993	0.922
2001–2005	0.958	0.988	0.947	0.996	0.993	0.936
2006–2010	0.964	0.991	0.956	0.997	0.994	0.947
2011–2015	0.971	0.996	0.967	0.997	0.995	0.960
Improvement (%)	*5.863*	*5.009*	*11.170*	*1.089*	*0.709*	*13.179*

Source Author's calculations

References

Claeson M, Bos ER, Mawji T, Pathmanathan I (2000) Reducing child mortality in India in the new millennium. Bull World Health Organ 78(10):11921199

Das Gupta P (1993) Standardization and decomposition of rates: a user's manual. Washington, DC. US Bureau of the Census, Current Population Reports, Series P23-186. US Government Printing Office

de Graft-Johnson J, Kerber K, Tinker A, Otchere S, Narayanan I, Shoo R, Oluwole D, Lawn J (2006) The maternal, newborn and child health continuum of care. In: Lawn J, Kerber K (eds) Opportunities for Africa's newborns. Cape town, Partnership for maternal, Newborn and child health, pp 23–36

De P, Sahu D, Pandey A, Gulati BK, Chandhiok N, Shukla AK, Mohan P, Mitra RG (2016) Post millennium development goals prospects on child mortality in India: an analysis using autoregressive integrated moving averages (ARIMA) model. Health 8:1845–1872

Durand JD (1948) The labour force in the United States: 1890–1960. Appendix B: methods of analyzing labour force change. Social Science Research Council, New York

Gill K, Pande R, Malhotra R (2007) Women deliver for development. Lancet 370:1347–1357
Government of India (1946) Report of the health survey and development committee. Manager of Publications, Delhi
Government of India (1974) National policy on children 1974. Department of Women and Child Development, New Delhi
Government of India (1975) The integrated child development scheme. Department of Women and Child Development, New Delhi
Government of India (1984) SRS based abridged life tables 1970-75. Registrar General of India. Occasional Paper No. 1, New Delhi
Government of India (1985) SRS based abridged life tables 1976-80. Registrar General of India. Occasional Paper No. 1, New Delhi
Government of India (1988) National mission on immunisation. Ministry of Health and Family Welfare, New Delhi
Government of India (1989) SRS based abridged life tables 1981-85. Registrar General of India. Occasional Paper No. 1, New Delhi
Government of India (1992) National child survival and safe motherhood programme. Ministry of Health and Family Welfare, New Delhi
Government of India (1994) SRS based abridged life tables 1986-90. Registrar General of India. Occasional Paper No. 1, New Delhi
Government of India (1997) Reproductive and child health programme. Scheme of implementation. Ministry of Health and Family Welfare, New Delhi
Government of India (2003) National technical advisory group on immunisation. Ministry of Health and Family Welfare, New Delhi
Government of India (2004) SRS based abridged life tables 1996-2000 and 1997. Registrar General of India. Occasional Paper No. 2, New Delhi
Government of India (2005) National rural health mission. Ministry of Health and Family Welfare, New Delhi
Government of India (2008) SRS based abridged life tables 2001-05. Registrar General of India, New Delhi
Government of India (2013) The national policy for children 2013. Ministry of Women and Child Development, New Delhi
Government of India (2015) The constitution of India. As on 9th November 2015. Ministry of Law and Justice, New Delhi
Government of India (2016) The Child Labour (prohibition and regulation) amendment Act 2016. Ministry of Law and Justice, New Delhi
Government of India (2016a) SRS Bulletin 50(2). Registrar General and Census Commissioner of India, New Delhi
Government of India (2017) SRS based abridged life tables. Ministry of Home Affairs, Office of the Registrar General and Census Commissioner, New Delhi
Government of India (2017a) The national health policy 2017. Ministry of Health and Family Welfare, New Delhi
Government of India (no date) National Family Health Survey - 4, 2015-16. India Fact Shhet. Ministry of Health and Family Welfare, New Delhi
Government of India (no date) Causes of death statistics 2007-13. Ministry of Home Affairs. Registrar General and Census Commissioner of India, New Delhi
Heligman I, Pollard JH (1980) The age pattern of mortality. J Inst Actuar (Oxford) Part I, 107(434): 49–80
Jain AK, Visaria P (1988) Infant mortality in india: differentials and determinants. Sage, New Delhi
Jones G, Schultnik W, Babille M (2006) Child survival in India. Indian J Pediatr 73:479–487
Kerber KJ, de Graft-Johnson JE, Bhutta ZA, Okong P, Starrs A, Lawn JE (2007) Continuum of care for maternal, newborn, and child health: from slogan to service delivery. Lancet 370:1358–1369
Kim YJ, Strobino DM (1984) Decomposition of the difference between two rates with hierarchical factors. Demography 21(3):361–372

Kuntla S, Goli S, Jain K (2014) Explaining sex differentials in child mortality in India: trends and determinants. Int J Popul Res. Article ID 649741
Lal RB, Swamy VS (1977) Vital statistics: sample registration system. In: Bose A, Gupta DB, Raychaudhuri G (eds) Population statistics in India. Vikas Publishing House, New Delhi
Lawn JE, Tinker A, Munjanja SP, Cousens S (2006) Where is maternal and child health now? Lancet 368:1474–1477
Measham A (1999) The performance of India and Indian states in reducing infant mortality and fertility 1975-90. Econ Polit Weekly 34(22):1359–1367
NIMS and UNICEF (2012) Infant and child mortality in india: trends and determinants. National Institute of Medical Statistics, New Delhi
NIPCCD (2014) An analysis of levels and trends in infant and child mortality rates in India. National Institute of Public Cooperation and Child Development, New Delhi
OECD (2005) The Paris declaration on aid electiveness. Organisation for Economic Cooperation and Development, Paris
Pandey A, Choe MK, Luther NY, Sahu D, Chand J (1998) Infant and child mortality in India. International Institute for Population Sciences. NFHS Subject Report No. 11, Mumbai
Puffer RR (1985) Mortality in infancy and childhood in India. United States Agency for International Development, New Delhi
Rogers A, Gard K (1991) Applications of the Heligman/Pollard model mortality schedule. Popul Bull United Nations 30:70–105
Selvin S (2004) Statistical analysis of epidemiological data. Oxford University Press, New York
Sokhey J, Kim-Farley RJ, Bhargava I (1989) The expanded programme of immunization. A decade of progress in India. Ann Trop Paediatr 9(1):24–29
Tilak JBG (1991) Socioeconomic correlates of infant mortality in India. The World Bank, Washington DC
Tinker A, Hoope-Bender P, Azfar S, Bustreo F, Bell R (2005) A continuum of care to save newborn lives. Lancet 365:822–825
Tyagi BN (1983) A review of diarrhoeal diseases control programme in India. Health Popul Perspect Issues 6(4):209–225
UNICEF (2015) Levels and trends in child mortality. Report 2015. Estimates Developed by UN Inter-Agency Group on Child Mortality Estimation. United Nations Children's Fund, New York
UNICEF (2016) The state of world's children 2016. A fair chance to every child. United Nations Children's Fund, New York
United Nations (1988) MortPak-Lite. The United Nations software package for mortality measurement. Interactive software for the IBM-PC and compatibles. United Nations, New York
United Nations (1990) United nations convention on the rights of the child. United Nations. Office of the High Commissioner on Human Rights, New York
United Nations (2015) Transforming our world. The 2030 agenda for sustainable development. United Nations, New York
Visaria L (1985) Infant mortality in India level, trends and determinants. Econ Polit Weekly 20(32):1352–1450
World Health Organisation (1986) The expanded programme on immunisation in South-East Asia. World Health Organisation. Regional Office for South-East Asia, New Delhi
World Health Organization (2005) World health report 2005: make every mother and child count. World Health Organization, Geneva, Switzerland

Chapter 6
Economic Growth, 1951–2012

Abstract This chapter analyses the growth of the Indian economy during the period 1950–1951 through 2011–2012 in the context of 11 Five-year Development Plans that the country had since independence. The analysis reveals that the acceleration in the growth of the economy has not been consistent during the period under reference so that the contribution of the growth acceleration to the growth of the real output of the economy has at best been marginal relative to the contribution of the increase in the size of the economy, or the momentum effects of economic growth. Moreover, population growth has been a big contributor to the economic growth. If the growth in the real output attributed to the increase in the population is excluded, then the growth of the real output of the economy has been highly sedate.

6.1 Introduction

The performance of the Indian economy in the context of long-term trend growth rate in the real gross domestic product at factor cost has been the subject of intensive empirical enquiry and debate in the past (Nagraj 1990). It is argued that the Indian economy grew at an almost constant rate growth of around 3.5% per year during the period 1950 through 1980 (Bardhan 1984) and was termed as the 'Hindu rate of growth', linking economic growth with Hindu culture (Virmani 2005). However, many authors expressed their reservation on the observation that the growth of the Indian economy stagnated till 1980 (Raj 1984; Patnaik 1987; Dhar 1988; Virmani 1989; Dandekar 1992; Ahluwalia 1995; Nagraj 1990). It is, however, widely believed that India's economy picked up only after the introduction of the new economic policy in 1991–1992 (Virmani 2005), although Virmani (1997a, b) has demonstrated that the growth acceleration preceded the new economic policy. In any case, it is well known that the growth of India's economy accelerated to more than 6% per year during 1990s and to almost 9% per year between 2003 and 2007 to become one of the fastest growing economies in the world after China. Even in the aftermath of the global financial crisis, the Indian economy recorded a healthy growth rate of around 6% during 2008–2009.

The primary driver behind economic growth in India since independence has been the Five-year Development Plans. Development planning in India has always been

© Springer Nature Singapore Pte Ltd. 2020

A. R. Chaurasia, *Population and Sustainable Development in India*,
https://doi.org/10.1007/978-981-32-9212-3_6

designed to lift up the Indian economy and improve the standard of living of the Indian people. Economic growth, therefore, received the topmost priority in all Five-year Development Plans that the country had since independence. The strategy adopted to accelerate economic growth has, however, been different in different Five-year Plans. The First Five-year Development Plan (1951–1956) focused primarily on the growth of the primary sector. The Second Plan (1956–1961), on the other hand, focused on the development of heavy industries with a dominant role of the public sector to spur economic growth. During the Third Plan (1961–1966), the emphasis was on making the economy independent and reaching self-active position for taking off. The Fourth Plan (1969–1974) focused on the growth of the economy with stability and progressive achievement of self-reliance. The Fifth Plan (1974–1979) gave top priority to agriculture followed by industry and mines. Poverty eradication and technological self-reliance was the emphasis of the Sixth Plan (1980–1985), whereas the Seventh Plan (1985–1990) was directed towards establishing self-sufficient economy and expanding opportunities for productive employment. The Eighth Plan (1992–1997) focused on the development of human resources while the Ninth Plan (1997–2002) emphasised growth with justice and equity. The Tenth Plan (2002–2007) aimed at doubling the per capita income in the next 10 years and the Eleventh Plan (2007–2012) focused on faster and more inclusive growth. The planned approach to development in the country had three discontinuities: 1966–1969; 1979–1980; and 1990–1992. The first discontinuity was the result of the dismal performance of the Third Five-year Plan and inflationary recession, while the second and third discontinuities were the result of political uncertainties (Government of India 2014).

This chapter analyses the growth of the Indian economy during the period 1950–1951 through 2011–2012 in the context of 11 Five-year Development Plans that the country had since independence. The analysis is based on the estimates of the output of the economy in real terms prepared and released by the Government of India, Ministry of Statistics and Programme Evaluation. The growth of the economy has been carried out in terms of the total output of the economy in real terms as well as in terms of the output in real terms of the nine main sectors of the economy. The primary purpose of the chapter is to explore how different Five-year Development Plans of the country had contributed to the growth of the economy of the country. Since accelerating economic growth has been one of the objectives of all Five-year Development Plans, the analysis contributes to the understanding of the relevance of development planning with reference to improving the quality of life of the Indian people.

The chapter is organised as follows: The next section describes the methodology adopted for the analysis. A regression-based approach has been used to analyse the growth of the monetary output of the economy. This approach involves application of a segmented or piecewise regression (Ryan and Porth 2007) to estimate the growth rate of the output of the economy during different Five-year Development Plans and during the three periods of plan discontinuity. The third section of the chapter describes the data used in the analysis. The analysis is based on the estimates of the real monetary output of the economy of the country that have been prepared by the Government of India. Results of the analysis are presented and discussed in section four of the chapter, while the last section discusses the policy relevance of the findings of the analysis.

6.2 Methodology

The monetary output of any production system can be measured in terms of either at the prevailing market prices or at the cost of factors of production (Nicholson 1955). Estimation of the monetary output in terms of market prices is sensitive to subsidies and indirect taxes, whereas estimation of the monetary output at factor cost is independent of subsidies and indirect taxes. It is, therefore, argued that the size and growth of the economy are better reflected when the monetary output of the economy is measured in terms of the cost of factors of production rather than in terms of prevailing market prices.

The indicator that is most commonly used to measure the monetary output of the economy is the domestic product which is defined as the total cost involved in producing all goods and services during a given time period within a specified administrative area without duplication. The domestic product can be measured in gross as well as in net terms. The net domestic product takes into account the cost of depreciation in the capital stock of the production system used in the production process, whereas the gross domestic product does not take into account the depreciation in the capital stock. The gap between the gross and the net domestic product reflects the obsolescence of the capital stock of the economy.

Estimation of the total cost involved in the production of goods and services follows a decentralised approach. First, the entire economy is divided into three main sectors of production—primary or agriculture sector, secondary or manufacturing sector and tertiary or service sector. Subsequently, each sector is further divided into a number of sub-sectors and each sub-sector is further divided into specific goods and services to create a nested system. The sum of the total cost of the production of goods and services by all sub-sectors and sectors then provides the estimate of the total output of the economy in monetary terms. If the total cost of the production is not adjusted for depreciation in the capital stock, it is termed as gross domestic product. If the total cost of the production is adjusted for the depreciation in the capital stock, it is termed as net domestic product.

The gross or the net domestic product may be estimated at the nominal (current) prices or at real (constant) prices. The gross domestic product (GDP) at nominal cost or current prices is a crude indicator of the monetary output of the economy because it is influenced by inflation or the increase in prices and it does not take into account the depreciation in the capital stock. The net domestic product (NDP) at fixed cost, on the other hand, is the most refined indicator of the monetary output of the economy. Estimation of NDP at either nominal or fixed cost is, however, problematic as it requires estimates of the depreciation of the capital stock, which is quite difficult to estimate correctly.

Costing of all goods and services produced by the economy is not the only method of estimating the domestic product. It is also possible to estimate the domestic product by what are known as the income approach and the expenditure approach. In principle, all the three approaches should give the same estimate of the domestic product but, in practice, this rarely happens. The expenditure approach is based on the principle that all goods and services produced must be purchased by somebody so that the total cost of goods and services produced must be equal to the total expenditure incurred in purchasing goods and services. The income approach, on the other hand, is based on the concept that total income of all producers of goods and services must be equal to the cost of all goods and services produced by them.

The Government of India follows a combination of production and income approaches to estimate the domestic product of the country. The details of the methodology adopted are given elsewhere (Government of India 2008). The estimation process involves first estimating the gross domestic product at the nominal cost or at current prices and then adjusting the gross domestic product at nominal cost for inflation. The net domestic product is then obtained by subtracting consumption of fixed capital from the gross domestic product. Estimates of the consumption of the fixed capital are prepared by using estimates of asset-wise net fixed capital stock and average life of the asset following the perpetual inventory method (Meinen et al. 1998).

The conventional approach to measure the growth of the economy is to calculate compound annual growth rate (CAGR). CAGR is based on the value of the domestic product in the beginning and at the end of the period under reference. It assumes a linear change in the domestic product within the period of estimation. CAGR is, therefore, best suited in situations where the reference period is small, preferably one year. When the reference period is long, CAGR may give a misleading picture of the growth of the economy as the growth may not be linear. In such a situation, one alternative is to break the reference period into smaller sub-periods and then calculate CAGR for each sub-period separately. This approach, however, treats different sub-periods independent of each other. It assumes that the growth in the economy in a sub-period is not influenced by the growth in earlier sub-periods. Such an assumption is not valid as the growth of the economy in a given period is influenced by the growth of the economy in earlier periods.

The alternative to CAGR is the piecewise or segmented regression approach (Ryan and Porth 2007), which implicitly recognises that growth in the economy in different time periods is different so that there are break points in the trajectory of the growth. For example, it may be assumed that the growth in the economy during the period 1950–1951 through 2011–2012 is essentially the result of the growth in the economy during different Five-year Development Plans. It may also be assumed that the growth of the economy is essentially linear within a plan period and that economic growth during different Five-year Development Plans has essentially been different. The use of segmented or piecewise regression ensures that growth of the economy during a given Five-year Development Plan takes into account the growth in the economy during the earlier Five-year Development Plans.

More specifically, the following piecewise regression model was fitted to the time series of the gross domestic product of the country for the period 1951–1952 (start of the First Five-year Development Plan) through 2011–2012 (last year of the XI Five-year Development Plan):

$$\text{In}(G) = a_0 + b_0 T + \sum_{i=2}^{13} b_i X_i + \epsilon \tag{6.1}$$

where G is the gross domestic product, T is the time with origin at 1951–1952, the first year of the First Five-year Development Plan and ε is the error term which is distributed normally. The variables X_1, .. X_{13}, on the other hand, are defined in the following manner:

X_1	$= 0$ if $T \leq$ 1956–1957 (Start of II Five-year Development Plan) $= T$-(1956–1957) if $T >$ 1956–1957
X_2	$= 0$ if $T \leq$ 1961–1962 (Start of III Five-year Development Plan) $= T$-(1961–1962) if $T >$ 1961–1962
X_3	$= 0$ if $T \leq$ 1966–1967 (Plan holiday 1966–1969) $= T$-(1966–67) if $T >$ 1966–1967
X_4	$= 0$ if $T \leq$ 1969–1970 (Start of IV Five-year Development Plan) $= T$-(1969–1970) if $T >$ 1969–1970
X_5	$= 0$ if $T \leq$ 1974–1975 (Start of V Five-year Development Plan) $= T$- (1974–1975) if $T >$ 1974–1975
X_6	$= 0$ if $T \leq$ 1979–1980 (Rolling Plan 1979–1980) $= T$-(1979–80) if $T >$ 1979–1980
X_7	$= 0$ if $T \leq$ 1980–1981 (Start of VI Five-year Development Plan) $= T$-(1980–1981) if $T >$ 1980–1981
X_8	$= 0$ if $T \leq$ 1985–1986 (Start of VII Five-year Development Plan) $= T$-(1985–1986) if $T >$ 1985–1986
X_9	$= 0$ if $T \leq$ 1990–1991 (Annual Plans 1990–1991 and 1991–1992) $= T$-(1990–1991) if $T >$ 1990–1991
X_{10}	$= 0$ if $T \leq$ 1992–1993 (Start of VIII Five-year Development Plan) $= T$-(1992–1993) if $T >$ 1992–1993
X_{11}	$= 0$ if $T \leq$ 1997–1998 (Start of IX Five-year Development Plan) $= T$-(1997–1998) if $T >$ 1997–1998
X_{12}	$= 0$ if $T \leq$ 2002–2003 (Start of X Five-year Development Plan) $= T$-(2002–2003) if $T >$ 2002–2003
X_{13}	$= 0$ if $T \leq$ 2007–2008 (Start of XI Five-year Development Plan) $= T$-(2007–2008) if $T >$ 2007–2008

The regression coefficients obtained on the basis of fitting the segmented or the piecewise regression model given by Eq. (6.1) have been used to estimate the average annual growth rate of the real gross domestic product during different Five-year Development Plans. The corresponding estimating formulas are as follows:

Growth rate during I Five-year Plan	$\exp(b_0) - 1$
Growth rate during II Five-year Plan	$\exp\left(\sum_{i=0}^{1} b_i\right) - 1$
Growth rate during III Five-year Plan	$\exp\left(\sum_{i=0}^{2} b_i\right) - 1$
Growth rate during Plan holiday 1966–1969	$\exp\left(\sum_{i=0}^{3} b_i\right) - 1$
Growth rate during IV Five-year Plan	$\exp\left(\sum_{i=0}^{4} b_i\right) - 1$
Growth rate during V Five-year Plan	$\exp\left(\sum_{i=0}^{5} b_i\right) - 1$
Growth rate during Rolling Plan 1979–1980	$\exp\left(\sum_{i=0}^{6} b_i\right) - 1$
Growth rate during VI Five-year Plan	$\exp\left(\sum_{i=0}^{7} b_i\right) - 1$
Growth rate during VII Five-year Plan	$\exp\left(\sum_{i=0}^{8} b_i\right) - 1$
Growth rate during Annual Plans 1990–1992	$\exp\left(\sum_{i=0}^{9} b_i\right) - 1$
Growth rate during VIII Five-year Plan	$\exp\left(\sum_{i=0}^{10} b_i\right) - 1$
Growth rate during IX Five-year Plan	$\exp\left(\sum_{i=0}^{11} b_i\right) - 1$
Growth rate during X Five-year Plan	$\exp\left(\sum_{i=0}^{12} b_i\right) - 1$
Growth rate during XI Five-year Plan	$\exp\left(\sum_{i=0}^{13} b_i\right) - 1$

The above specification shows that the growth of the economy during a given Five-year Development Plan is determined by the growth of the economy during the period prior to it. For example, the growth of the economy during the XI Five-year Development Plan (2007–2012) is determined by the growth of the economy during all Five-year Development Plans prior to 2007 and during plan holiday period, rolling plan period and during annual plans as described above. Results of the piecewise regression analysis make it possible to estimate how much of the growth in the economy during a given period is attributed to the growth in the past and the growth during the current period. The growth of the economy during the past has a momentum or size effect on the growth of the economy during the current period which when multiplied by the tempo of growth during the current period determines the growth of the economy during the current period.

6.3 Data Source

The analysis is based on the annual estimates of the gross domestic product (GDP) prepared by the Government of India, Ministry of Statistics and Programme Implementation for the period 1951–1952, the beginning of the I Five-year Development Plan through 2011–2012, the last year of the XI Five-year Development Plan. These estimates are available at both current prices and at fixed (2004–2005) prices as well as at factor cost and at market prices. The present analysis is based on the estimates of the GDP at factor cost at fixed (2004–2005) prices. The Government of India has also prepared and released estimates of net domestic product (NDP) which is obtained from the GDP after accounting for the consumption of the fixed capital in the production process. However, estimates of NDP have not been used in the present analysis because of the poor reliability of the estimates of the consumption of the fixed capital. These estimates are also available for the 19 sub-sectors of the economy which have been further grouped into the following nine main sub-sectors:

1. Agriculture, forestry and fishing
2. Mining and querying
3. Manufacturing
4. Electricity, gas and water supply
5. Construction
6. Trade, hotels and restaurant
7. Transport, storage and communication
8. Financing, insurance, real estate and business services
9. Community, social and personal services.

6.4 Economic Growth in India

The GDP at factor cost in India was estimated to be Rs 2991.27 billion in 1951–1952 at 2004–2005 prices which increased to Rs 52,475.30 billion in 2011–2012. If the increase in real GDP is any indication then India's economy expanded by more than 17.5 times between 1951–1952 and 2011–2012 (Table 6.1). During the same period, the real output of the primary or the agriculture sector increased by around five times from Rs 1675.63 billion in 1951–1952 to Rs 8645.57 billion in 2011–2012; that of the secondary or manufacturing sector increased by almost 31 times from Rs 275.87 billion to Rs 8540.98 billion; and that of the tertiary or service sector increased by almost 34 times from Rs 1039.77 billion in 1951–1952 to Rs 35,288.75 billion in 2011–2012 (Fig. 6.1). This means that almost 70% of the increase in the real output of the economy of the country during the period under reference was confined to the services sector, whereas the contribution of the agriculture sector was less than 15% while that of the manufacturing sector was slightly more than 15%.

Among the nine main sub-sectors of the economy, the increase in the real output during the period under reference was not the same (Fig. 6.2). The contribution of the increase in the real output of financing, insurance, real estate and business services to the increase in the real GDP was the highest (18.6%), while that of electricity, gas and water supply sub-sector was the lowest (2.02%). The increase in the real output of the manufacturing sector and trade, hotel and restaurant sub-sector to the increase in real GDP was also more than 15%, whereas the contribution of the increase in the real output of mining and quarrying sub-sector was just around 2%. The increase in the real output of the construction sub-sector also accounted for just around 8% of the increase in the real output of the economy of the country during this period.

Table 6.1 Increase in the real GDP in India, 1951–2012 (billion rupees)

Sector/Sub-sector		Real output during		Increase in real output	Contribution to the increase in real output
		1951–1952	2011–2012		
Real GDP		2991.27	52475.30	49484.00	100.00
Primary		1675.63	8645.57	6969.94	14.09
I	Agriculture	1617.91	7538.32	5920.41	11.96
II	Mining and quarrying	57.72	1107.25	1049.53	2.12
Secondary		275.87	8540.98	8265.11	16.70
III	Manufacturing	275.87	8540.98	8265.11	16.70
Tertiary		1039.77	35288.80	34249.00	69.21
IV	Electricity, gas and water supply	8.83	1006.46	997.63	2.02
V	Construction	152.25	4151.88	3999.63	8.08
VI	Trade, hotel and restaurant	240.92	8466.30	8225.38	16.62
VII	Transport, storage and communication	90.15	5556.31	5466.16	11.05
VIII	Financing, insurance, real estate and business services	243.18	9455.34	9212.16	18.62
IX	Community, social and personal services	304.44	6652.46	6348.02	12.83

Source Author's calculations based on Appendix Table A.6.1

Results of the analysis of the trend in the real GDP and in the real output of the nine main sub-sectors of the economy of the country, based on the application of the segmented or piecewise regression model described by the Eq. (6.1) to the data given in the Appendix Table A.6.1, are presented in Table 6.2. The piecewise regression model has provided an excellent fit to the increase in the real GDP as well as to the increase in the real output of different sub-sectors of the economy during the period under reference. In case of real GDP, the piecewise regression model could explain more than 99% of the variation in the real GDP over time. Similarly, the regression model accounted for more than 99% of the variation in the real output of all the nine main sub-sectors of the economy.

Results of the regression model have been used to estimate the average annual growth rate of real GDP during different Five-year Development Plans. The average annual growth rates in different Five-year Development Plans for the total output (GDP) and for outputs of different sub-sectors of the economy are presented in Table 6.3. The Indian economy expanded most rapidly during X Five-year Develop-

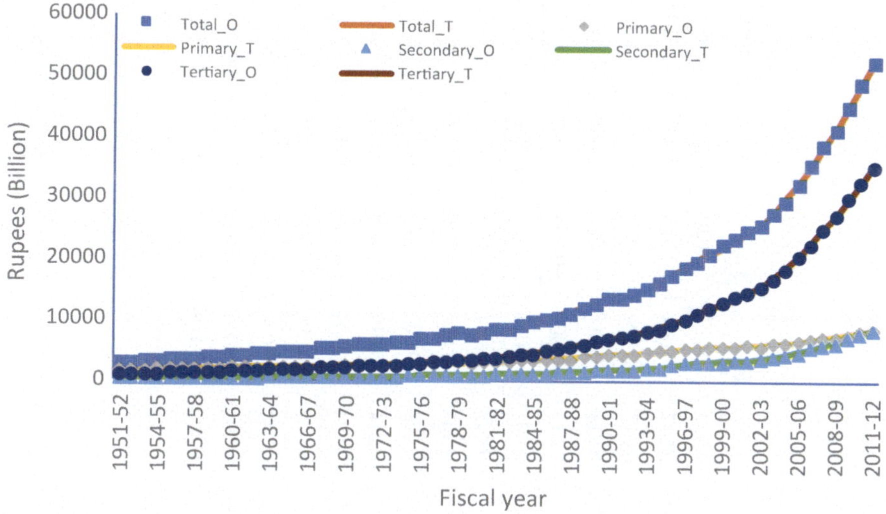

Fig. 6.1 Trend in gross domestic product at 2004–2005 prices in India

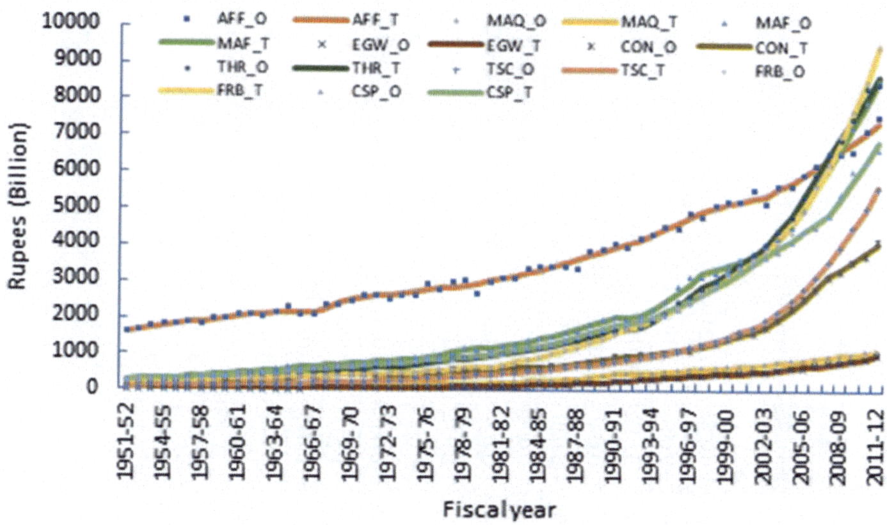

Fig. 6.2 Trend in the out of different sectors of Indian economy at 2004–2005 prices

Table 6.2 Results of the application of the piecewise regression model

Parameters of the model	GDP	A	Q	M	E	C	H	T	F	S	Primary	Tertiary
b0	0.035	0.025	0.024	0.060	0.085	0.068	0.051	0.053	0.029	0.030	0.025	0.043
b1	0.006	−0.001	0.049	0.007	0.043	−0.005	0.010	0.017	0.004	0.016	0.001	0.010
b2	−0.008	−0.018	−0.013	−0.018	−0.017	0.018	−0.012	−0.017	−0.004	0.009	−0.017	0.001
b3	0.012	0.042	−0.051	−0.014	−0.012	−0.037	−0.015	−0.011	0.011	−0.011	0.038	−0.012
b4	−0.017	−0.030	0.019	0.004	−0.044	−0.047	0.001	0.013	−0.006	−0.006	−0.028	−0.010
b5	0.011	−0.006	0.020	0.018	0.028	0.057	0.023	0.007	0.018	−0.003	−0.004	0.020
b6	−0.015	0.023	0.086	−0.070	−0.073	−0.026	−0.058	−0.007	−0.052	−0.009	0.027	−0.034
b7	0.022	−0.012	−0.076	0.066	0.069	−0.012	0.054	0.003	0.086	0.029	−0.016	0.039
b8	0.013	0.008	0.031	0.007	0.006	0.052	0.004	0.002	0.005	0.012	0.011	0.014
b9	−0.027	−0.007	−0.071	−0.051	−0.013	−0.041	−0.032	−0.014	−0.017	−0.049	−0.012	−0.031
b10	0.036	0.011	0.031	0.079	0	0.020	0.069	0.029	0.004	0.049	0.012	0.036
b11	−0.016	−0.019	−0.012	−0.054	−0.038	0.023	−0.030	0.005	−0.009	−0.001	−0.018	−0.008
b12	0.031	0.017	0.010	0.059	0.034	0.046	0.031	0.042	0.029	−0.013	0.016	0.024
b13	−0.002	0.005	−0.009	−0.012	−0.009	−0.055	−0.028	−0.008	0.012	0.031	0.003	−0.004
Intercept	12.605	12.002	8.659	10.220	6.754	9.537	10.077	9.078	10.103	10.318	12.037	11.532
R^2	0.9995	0.9936	0.999	0.9992	0.9998	0.9975	0.9996	0.9998	0.9997	0.9995	0.9953	0.9998

Source Author's calculations

Table 6.3 Average annual growth rate of real GDP and real output of different sectors of the economy in India during different plan periods

Plan	GDP	AGR	MAN	SER	AFF	MAQ	MAF	EGW	CON	THR	TSC	FRB	CSP	CV
I	3.542	2.559	6.145	4.390	2.565	2.475	6.145	8.841	7.038	5.214	5.396	2.952	3.075	0.437
II	4.114	2.646	6.905	5.406	2.452	7.671	6.905	13.628	6.463	6.251	7.242	3.336	4.737	0.464
III	3.240	0.907	5.000	5.527	0.637	6.251	5.000	11.737	8.427	5.011	5.401	2.923	5.647	0.521
A1	4.441	4.776	3.543	4.270	4.983	0.963	3.543	10.459	4.484	3.471	4.264	4.050	4.537	0.523
IV	2.710	1.898	3.911	3.258	1.846	2.940	3.911	5.740	−0.300	3.619	5.659	3.428	3.926	0.511
V	3.795	1.463	5.795	5.302	1.252	4.987	5.795	8.718	5.586	6.073	6.367	5.255	3.662	0.361
A2	2.214	4.257	−1.317	1.810	3.562	14.433	−1.317	1.114	2.894	0.142	5.603	−0.074	2.704	1.374
VI	4.539	2.604	5.393	5.808	2.343	6.061	5.393	8.387	1.691	5.687	5.933	8.904	5.700	0.401
VII	5.883	3.759	6.181	7.279	3.196	9.455	6.181	9.093	7.122	6.091	6.195	9.404	6.926	0.270
A3	3.080	2.486	0.936	4.044	2.512	1.982	0.936	7.720	2.824	2.764	4.751	7.546	1.861	0.639
VIII	6.818	3.763	9.280	7.854	3.599	5.158	9.280	7.742	4.883	10.113	7.814	8.007	7.014	0.285
IX	5.143	1.937	3.549	7.044	1.678	3.871	3.549	3.674	7.292	6.910	8.314	7.045	6.961	0.395
X	8.402	3.569	9.886	9.695	3.376	4.939	9.886	7.242	12.363	10.311	12.958	10.178	5.551	0.375

(continued)

Table 6.3 (continued)

Plan	GDP	AGR	MAN	SER	AFF	MAQ	MAF	EGW	CON	THR	TSC	FRB	CSP	CV
XI	8.177	3.870	8.588	9.243	3.853	3.959	8.588	6.270	6.313	7.318	12.042	11.499	8.829	0.361
CV	0.399	0.386	0.589	0.374	0.423	0.600	0.589	0.402	0.579	0.476	0.358	0.521	0.356	0.545

Source Author's calculations
Remarks GDP Gross Domestic Product (Real prices)
AGE Agriculture sector
MAF Manufacturing sector
SER Service sector
AFF Agriculture, forestry and fishing
MAQ Mining and querying
MAF Manufacturing
EGW Electricity, gas and water supply
CON Construction
THR Trade, hotels and restaurant
TSC Transport, storage and communication
FRB Financial, insurance, real estate and business services
CSP Community, social and personal services

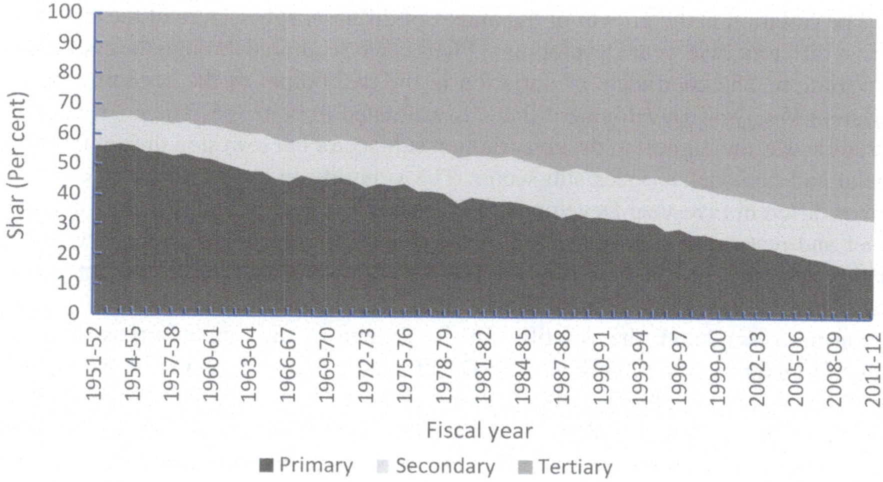

Fig. 6.3 Components of growth in the real output of different sectors of Indian economy

ment Plan (2002–2007) when the real GDP increased at more than 8.4% per year. In contrast, the expansion was the slowest during IV Five-year Plan (1969–1974) when the real GDP increased at just around 2.7% per year. Figure 6.3 suggests that the economy of the country picked up only after the launch of VI Five-year Plan in 1980–1981. Before 1980–1981, the average annual growth rate of the economy of the country never crossed the 5% mark.

The growth of the real output of different sub-sectors of the economy varied widely across different Five-year Development Plans (Table 6.3). The growth of the output of the agriculture, forestry and fishing sub-sector was most rapid during XI Five-year Development Plan but least rapid during III Five-year Development Plan. The output of the mining and quarrying sub-sector, on the other hand, grew most rapidly during VII Five-year Development Plan but least rapidly during I Five-year Development Plan. The growth of the output of the manufacturing sub-sector was most rapid during X Five-year Development Plan but least rapid during IX Five-year Development Plan. The growth of the real output of electricity, gas and water supply sector was most rapid during II Five-year Plan but least rapid during IX Five-year Development Plan, whereas the growth of the output of construction, trade hotel and restaurant, and transport, storage and communication sub-sectors was most rapid during X Five-year Development Plan but least rapid during IV Five-year Development Plan. Finally, the growth of the output of financial, insurance, real estate and business services sub-sector; and community, social and personal services sub-sector was most rapid during the XI Five-year Development Plan. However, the growth of the real output of financial, insurance, real estate and business services sub-sector was least rapid during the III Five-year Development Plan whereas the growth of the real output of community, social and personal services sub-sector was least rapid during the I five-year Development Plan.

The variation in the growth of the output of different sub-sectors of the economy across different Five-year Development Plans can be captured through the coefficient of variation. The coefficient of variation in the real output of the economy across different Five-year Development Plans is estimated to be 0.365. The coefficient of variation was the highest in the construction sub-sector but lowest in the community, social and personal services sub-sector. The volatility in the growth of the output across different Five-year Development Plans has also been found to be low in trade, hotel and restaurant sub-sector and in the manufacturing or the secondary sector of the economy. A low volatility implies higher consistency in the output across different Five-year Development Plans.

On the other hand, the volatility in the growth of the output across different sub-sectors during the same Five-year Development Plan is found to be the highest in III Five-year Development Plan, followed by IV Five-year Development Plan but the lowest in VII and VIII Five-year Development Plans. A low volatility in the growth of the output across different sub-sectors suggests consistency in the expansion of the economy across its different sectors and sub-sectors. This means that during VII and VIII Five-year Development Plans, the expansion of the economy was more consistent across different sectors and sub-sectors than other plans. In contrast, the expansion of the economy during III and IV Five-year Development Plans was largely confined to the electricity, gas and water supply sub-sector. The expansion of the economy was also quite rapid in the construction sector during III Five-year Development Plan but in transport, storage and communication sector during IV Five-year Development Plan.

The difference in the average annual rate of growth of the economy between the Five-year Development Plans may be attributed to two factors. The first factor is the size of the economy at the beginning of the plan, while the second is the tempo of the growth during the Five-year Development Plan. For example, the difference between the average annual growth rate during the I Five-year Development Plan (g_I) and the II Five-year Development Plan (g_{II}) may be written as:

$$g_{II} - g_I = \left[e^{(b_2+b_1)} - 1\right] - \left[e^{b_0} - 1\right] = e^{b_0}(e^{b_1} - 1) \tag{6.2}$$

In other words, the difference in the average annual growth rate of the economy between two Five-year Development Plans is determined by two factors. The first factor reflects the ratio of the size of the economy at the start of the two Development Plans and, therefore, may be termed as the size component while the second factor reflects the tempo of the growth in the second (later) Development Plan. The tempo of growth reflects the acceleration or deceleration in the growth in the second (later) Development Plan relative to the first (earlier) Development Plan. When the tempo of growth is more than 1, the growth in the second (later) Development Plan is faster relative to the growth in the first (earlier) Development Plan which indicates growth acceleration. When the tempo of growth is less than 1, the growth in the second (later) Development Plan is slower relative to the growth in the first (earlier) Development Plan which indicates growth deceleration. When the tempo of growth is equal to 1, there is neither acceleration nor deceleration in the growth of the economy during the second (later) Development Plan relative to the growth during the first (earlier) Development Plan.

Table 6.4 presents the results of the decomposition of the difference between the average annual growth rates of successive Five-year Development Plans. During the I Five-year Development Plan, the economy of the country grew by 1.03542 times every year so that the real output of the economy increased from Rs 2981.15 billion in 1951–1952 to Rs 3426.54 billion in 1955–1956. During the II Five-year Development Plan, the growth of the economy accelerated so that the real output of the economy of the country grew by 1.00552 times every year. As a result, the real output of the economy increased to Rs 4168.84 billion during 1960–1961. Had the tempo of growth during the II Five-year Development Plan would have been the same as during the I Five-year Development Plan, the real output of the economy would have increased to 4077.99 billion Rupees or around 90.85 billion Rupees less in real terms. During the III Five-year Development Plan, however, the tempo of growth decelerated leading to the decrease in the real output of the economy during the plan. During the III Five-year Plan, the real output of the economy grew by 0.9916 times every year so as to reach 4930.78 billion Rupees by 1965–1966. Had the tempo of growth not decelerated during the III Five-year Development Plan, the economy of the country would have grown to 5099.96 billion Rupees by 1965–1966. This means that the deceleration of growth during the III Five-year Development Plan compared

Table 6.4 Components of growth in real output in different Five-year Development Plans in India

Plan period	Average annual growth rate	Difference in average annual growth rate	Multipliers of the difference in growth rate		Increase in real output (billion rupees) attributed to		
			Size multiplier	Tempo multiplier	Increase in size	Growth accelera-tion	Total
I	0.0354				445.39		445.39
II	0.0411	0.0057	1.03542	0.00552	651.45	90.85	742.30
III	0.0324	−0.0087	1.04114	−0.0084	931.12	−169.18	761.94
A1	0.0444	0.012	1.03240	0.01163	494.95	126.97	621.91
IV	0.0271	−0.0173	1.04441	−0.01658	1347.41	−446.23	901.17
V	0.0379	0.0109	1.02710	0.01056	923.10	316.63	1239.72
A2	0.0221	−0.0158	1.03795	−0.01523	594.95	−126.23	468.72
VI	0.0454	0.0233	1.02214	0.02275	747.14	838.66	1585.79
VII	0.0588	0.0134	1.04539	0.01286	2422.27	638.17	3060.44
A3	0.0308	−0.0280	1.05883	−0.02647	1551.35	−380.10	1171.25
VIII	0.0682	0.0374	1.03080	0.03626	2289.81	2491.06	4780.87
IX	0.0514	−0.0168	1.06818	−0.01568	7328.70	−1598.14	5730.56
X	0.0840	0.0326	1.05143	0.03100	6979.71	4087.00	11066.71
XI	0.0818	−0.0023	1.08402	−0.00210	17667.50	−439.76	17227.74
All					44374.83	5429.68	49804.51

Source Author's calculations

to the growth during II Five-year Development Plan had resulted in a decrease of more than 169 billion Rupees in the real output of the economy during III Five-year Development Plan. Similarly, the growth in the economy accelerated during the plan holiday period (1966–1967 to 1968–1969) relative to the growth during the III Five-year Development Plan. This acceleration in the growth resulted in an addition of almost 127 billion Rupees to the real output during this period. Similarly, the growth during the XI Five-year Development Plan decelerated relative to the growth during the X Five-year Development Plan. This deceleration resulted in a decrease of almost 440 billion Rupees in the real output of the economy during XI Five-year Development Plan. In any case, it is clear from Table 6.4 that the growth in the Indian economy frequently accelerated or decelerated in successive Five-year Plans which indicates towards a lack of consistency in the growth history of the Indian economy. There was growth acceleration in the II Five-year Development Plan relative to I Five-year Development Plan but the growth decelerated during the III Five-year Development Plan relative to II Five-year Development Plan. However, the growth accelerated again during the plan holiday period relative to III Five-year Development Plan but decelerated during IV Five-year Development Plan relative to the growth during the plan holiday period. In fact, growth acceleration in a Five-year Development Plan has always been followed by a growth deceleration in the next Five-year Development Plan with the only exception of VI and VII Five-year Development Plans. As a result, the contribution of the growth acceleration to the increase in the real output of the economy of the country during 1951–1952 through 2011–1912 is estimated to be only around 10%. Rest of the increase in the real output of the economy may be attributed to the increase in the size of the economy. This shows that the economic growth in the country during the period under reference has largely been the result of the momentum of growth built in the size of the economy rather than growth acceleration. This is a negative, yet, an interesting aspect of economic growth in India. Had India been able to consistently accelerate growth in successive Five-year Development Plans, the story of economic growth in India would have been radically different. For example, if the targets set for economic growth under different Five-year Development Plans would have been met, the real output of the Indian economy would have increased to Rs 78,197.31 billion during 2011–2012 against the estimated Rs 52,785.67 billion. In other words, the economy of the country would have expanded by more than 26 times against the current around 18 times.

The gross domestic product (GDP) may also be conceptualised as the sum of the total final consumption expenditure, gross capital formation and the balance of exports and imports (OECD/Eurostat 2012). This means that growth of the final consumption expenditure has a strong impact on the growth of GDP. Since growth of the final consumption expenditure is directly related to the growth of the population, it is logical to argue that population growth has a strong impact on the growth of GDP. In other words, a part of the growth in real GDP may be attributed to the increase in population. In fact, the increase in the real GDP in India during 1950–1951 through 2011–2012 has been found to be almost linearly correlated with the increase in population during this period with a simple zero-order correlation coefficient of

0.9190. Similarly, the simple zero-order correlation coefficient between the growth of the real output of the primary sector and the increase in population is estimated to be 0.9838. On the other hand, the simple zero-order correlation coefficient between the growth of the real output of the secondary sector with the increase in population is estimated to be 0.9166 and the simple zero-order correlation coefficient between the growth of the real output of the tertiary sector and population increase is estimated to be 0.8975. These correlation coefficients suggest an almost linear direct relationship between the growth of the real output of the economy and the increase in population during the period under reference. This implies that a substantial proportion of the growth of the real output of the economy of the country was the result of the increase in population from around 361 million in 1951 to almost 1211 million in 2011.

An idea about the contribution of the increase in the population to the growth of GDP may be made by decomposing the growth of GDP into the increase in the population and the increase in the output per capita. If Y denotes the real GDP, I denotes the real output per capita and P denotes the population, then the increase in the real GDP over time can be decomposed as

$$\frac{Y_t}{Y_0} = \frac{I_t * P_t}{I_0 * P_0} = \frac{I_t}{I_0} * \frac{P_t}{P_0} \tag{6.2}$$

Application of the above formulation to India suggests that the real GDP in India increased by around 17.5 times between 1951–1952 and 2011–2012 while the population increased by around 3.4 times during this period. This means that real output per capita increased by around 5.2 times between 1951–1952 and 2011–2012. This observation again confirms that a substantial proportion of the growth in real GDP in the country during the period under reference has been due to the increase in the population of the country.

Table 6.5 decomposes the growth in the real output of the economy of the country during 1951–1952 through 2011–2012 into the growth attributed to the increase in population and the growth attributed to the increase in per capita real output of the economy as a whole as well as for different sub-sectors of the economy following the Kitagawa decomposition (Kitagawa 1955). The decomposition exercise suggests that around 44% of the increase in the real output of the economy during 1951–1952 through 2011–2012 may be attributed to the increase in population, whereas the increase in the real output per capita accounted for around 56% of the increase. In the agriculture, forestry and fishing sub-sector of the economy, more than three-fourths of the increase in the real output was the result of the increase in population. Although the contribution of the increase in population to the growth of the real output in different sub-sectors varies, yet, it is clear that the increase in population had a substantial impact on the growth of the real output in all sub-sectors of the economy. This means that a very substantial proportion of the observed growth in the real output of the economy of the country during 1951–2012 was due to the inflammatory effect of the increase in population and not due to any real expansion of the economy.

Table 6.5 Decomposition of the increase in real output of the economy

Sectors of economy		Increase in real output (billion rupees)	Increase in real output attributed to the increase in		Proportionate contribution	
			Real output per capita	Population	Real output per capita	Population
All		49484.030	27550.940	21933.090	55.68	44.32
I	Primary	6969.938	1964.573	5005.365	28.19	71.81
I.1	Agriculture, forestry and fishing	5920.405	1371.483	4548.922	23.17	76.83
I.2	Mining and quarrying	1049.533	593.090	456.443	56.51	43.49
II	Secondary (Manufacturing)	8265.110	4943.511	3321.600	59.81	40.19
III	Tertiary	34248.980	20642.860	13606.130	60.27	39.73
III.4	Electricity, gas and water supply	997.629	634.077	363.553	63.56	36.44
III.5	Construction	3999.633	2363.612	1636.021	59.10	40.90
III.6	Trade, hotels and restaurant	8225.379	4971.107	3254.272	60.44	39.56
III.7	Transport, storage and communication	5466.159	3410.397	2055.762	62.39	37.61
III.8	Financial, insurance, real estate and business services	9212.163	5608.187	3603.976	60.88	39.12
III.9	Community, social and personal services	6348.021	3655.478	2692.543	57.58	42.42

Source Author's calculations

6.5 Transition in the Structure of the Economy

Expansion of the economy of the country has been associated with significant changes in the structure of the social and economic production system. During 1951–1952, the output of the primary sector accounted for more than 56% of the total output of the economy in real terms. This proportion decreased to just around 16% during 2011–2012, suggesting that the economy of the country is now hardly sensitive to the vagaries of the primary sector of the economy (Fig. 6.3). It is well known that the output of the primary sector of the economy, especially, output of the agriculture sub-sector, is dependent to a very substantial extent on the nature, particularly rains as most of the agriculture in the country is traditional and depends upon rains, and, therefore, the output of the primary sector, especially, agriculture remains contingent upon the nature.

At the same time, the share of the output of the secondary sector increased from around 9% to more than 16%, whereas the share of the tertiary sector increased from around 35% to more than 67% during this period. Alternatively, the real output of the economy of the country increased by around 49,484 billion Rupees between 1951–1952 and 2011–2012. Around 14% of this increase was confined to the primary sector, another 17% to the secondary sector and the remaining 69% to the tertiary sector. Within the tertiary sector, growth of the real output of financial, insurance, real estate and business services accounted for almost 19% of the growth of the real output of the economy, whereas growth of the output of trade, hotel and restaurant sub-sector accounted for almost 17% of the increase in the total output. In contrast, growth of the output of electricity, gas and water supply sub-sector accounted for just around 2% of the increase in the total output of the economy. Similarly, the growth of the real output of the construction sub-sector accounted for just around 8% of the increase in the total output. It is also evident from Fig. 6.3 that the decrease in the share of the primary sector has been associated with an almost equal increase in the share of the tertiary sector to the total output of the economy but the share of the secondary sector increased only marginally.

6.6 Conclusions

The present analysis has presented the historical perspective of the evolution of the economy of the country in terms of the growth of the real output of the economy at factor cost and in the context of 11 Five-year Development Plans that the country had since 1951–1952. A segmented regression model was applied to measure the growth in the output of the economy during different Five-year Development Plans which provided an excellent fit to the observed data on the real output of the economy and the real output of different sectors and sub-sectors of the economy. The analysis has revealed that the growth and expansion of the economy of the country have primarily been driven by the growth and expansion of the tertiary or the service sector of the economy during the period under reference. The analysis also reveals that there has been limited expansion of the primary or the agriculture sector.

There are three main conclusions that can be drawn from the analysis presented in this chapter. First, the acceleration in the growth of the economy has not been consistent during the period under reference. The acceleration in the growth of the real output of the economy during a Five-year Development Plan has almost invariably been associated with the deceleration in the growth of the real output in the successive Five-year Development Plan. Second, as a result of inconsistent growth acceleration across different Five-year Development Plans, the contribution of the growth acceleration to the growth of the real output of the economy has, at best, been marginal relative to the contribution of the increase in the size of the economy, or the momentum effects of economic growth. An acceleration in the growth of the output of the economy reflects an increase in the productivity of the social and economic production system. This means that there has been frequent ups and downs in the productivity of the social and economic production system of the country.

Third, population growth has been a big contributor to the economic growth that the country has witnessed in the past. If the growth in the real output attributed to the increase in the population is excluded then the growth of the real output of the economy has been highly sedate, which reflects an unsatisfactory performance of the social and economic production system of the country. The analysis suggests that without substantially increasing the productivity of the social and economic production system of the country, efforts to curb population will have a negative impact on the growth of the economy.

Appendix

See Appendix Table A.6.1

Table A 6.1 Estimates of GDP (billion rupees) in India at 2004–2005 prices; 1951–2012

Year	Total GDP	Agriculture, forestry and fishing	Mining and querying	Manufacturing	Electricity, gas and water supply	Construction	Trade, hostels and restaurants	Transport, storage and communi-cation	Financing, insurance, real estate and business services	Community, social and personal services
1951–1952	2991.3	1617.9	57.7	275.9	8.8	152.2	240.9	90.2	243.2	304.4
1952–1953	3048.7	1647.4	59.0	286.8	9.2	141.2	249.2	92.0	253.0	310.6
1953–1954	3204.2	1747.4	59.9	310.6	10.0	145.6	258.6	95.1	257.0	320.1
1954–1955	3342.2	1801.2	62.5	330.4	10.8	163.6	276.3	99.8	266.2	331.4
1955–1956	3444.7	1798.8	63.5	353.9	12.0	194.6	296.4	107.4	276.7	341.4
1956–1957	3626.0	1882.0	66.7	378.3	13.1	216.9	317.1	115.9	281.6	354.4
1957–1958	3588.8	1810.7	71.0	392.3	15.1	190.4	323.3	123.8	292.0	370.1
1958–1959	3836.5	1966.2	73.3	413.2	17.1	212.8	336.4	132.4	300.2	385.1
1959–1960	3931.5	1958.2	77.0	438.9	19.7	227.4	357.0	140.6	311.4	401.4
1960–1961	4198.7	2075.3	88.6	472.5	21.3	263.0	390.1	148.7	318.6	420.6
1961–1962	4335.4	2087.8	93.7	510.7	24.3	272.2	415.0	159.4	332.1	440.1
1962–1963	4434.5	2051.7	104.8	545.5	27.3	282.3	436.9	171.5	343.6	470.9
1963–1964	4669.1	2110.6	107.9	595.1	32.3	316.8	468.9	181.8	354.5	501.3
1964–1965	4987.7	2273.7	109.4	634.6	35.2	342.3	503.3	190.6	364.8	533.9
1965–1966	4896.4	2094.5	122.3	637.5	38.9	365.1	506.6	200.6	376.1	554.8
1966–1967	4981.2	2086.3	125.2	643.4	42.3	395.1	519.0	206.5	383.4	580.0
1967–1968	5346.5	2342.3	129.0	650.8	47.0	423.5	537.9	219.1	394.3	602.6
1968–1969	5485.2	2342.0	132.6	685.3	53.1	438.3	561.2	230.6	412.9	629.2

(continued)

Table A 6.1 (continued)

Year	Total GDP	Agriculture, forestry and fishing	Mining and querying	Manufacturing	Electricity, gas and water supply	Construction	Trade, hostels and restaurants	Transport, storage and communi-cation	Financing, insurance, real estate and business services	Community, social and personal services
1969–1970	5795.7	2468.9	139.2	749.6	57.8	452.0	591.3	243.5	429.9	663.6
1970–1971	6064.9	2635.3	129.7	766.1	61.4	451.0	623.1	251.2	446.8	700.5
1971–1972	6150.9	2610.1	133.1	793.5	66.4	452.9	635.1	259.0	469.4	731.3
1972–1973	6154.4	2496.7	140.9	825.8	69.5	463.6	637.2	278.3	487.2	755.2
1973–1974	6374.4	2638.6	142.7	861.8	71.0	433.5	664.5	288.1	499.5	774.7
1974–1975	6493.0	2637.6	149.8	890.4	74.2	419.7	695.3	314.6	499.8	811.7
1975–1976	7056.7	2936.2	168.0	911.1	85.0	479.5	761.4	343.0	532.7	839.9
1976–1977	7146.7	2777.1	173.9	984.2	94.7	526.4	788.3	366.7	572.6	862.8
1977–1978	7586.9	2969.7	179.4	1044.4	99.3	579.7	853.6	375.5	599.8	885.5
1978–1979	7979.1	3049.6	184.2	1176.6	110.6	566.7	927.9	399.8	640.8	922.9
1979–1980	7607.7	2678.1	186.2	1135.9	111.9	536.8	896.9	422.3	649.0	990.5
1980–1981	8120.9	2972.6	208.9	1142.5	118.4	607.6	944.2	451.9	662.6	1012.2
1981–1982	8577.4	3103.0	237.5	1235.4	129.6	640.8	997.8	482.8	716.7	1033.8
1982–1983	8820.5	3088.5	265.7	1271.3	138.1	595.8	1058.7	503.8	783.5	1115.1
1983–1984	9463.2	3357.2	273.4	1396.1	147.7	628.0	1113.3	530.3	860.5	1156.8
1984–1985	9834.1	3407.2	276.6	1450.6	163.7	649.8	1161.9	561.0	924.7	1238.6
1985–1986	10255.5	3419.3	291.7	1497.4	176.7	686.5	1256.7	603.5	1012.9	1310.8
1986–1987	10708.5	3402.6	327.4	1580.5	194.9	703.0	1328.8	642.7	1117.7	1411.0

(continued)

Table A 6.1 (continued)

Year	Total GDP	Agriculture, forestry and fishing	Mining and querying	Manufacturing	Electricity, gas and water supply	Construction	Trade, hostels and restaurants	Transport, storage and communi-cation	Financing, insurance, real estate and business services	Community, social and personal services
1987–1988	11100.1	3348.4	339.7	1667.4	210.0	743.3	1387.0	688.8	1200.3	1515.1
1988–1989	12168.4	3819.4	394.7	1804.2	230.4	795.5	1479.2	721.0	1316.5	1607.6
1989–1990	12952.5	3894.1	424.7	1958.1	252.8	851.5	1596.2	767.3	1472.2	1735.7
1990–1991	13634.7	4032.0	469.1	2050.0	269.7	951.8	1679.8	804.1	1567.7	1810.4
1991–1992	13847.2	3963.7	484.8	1998.5	295.9	971.4	1689.4	853.9	1732.5	1857.0
1992–1993	14551.7	4196.5	489.3	2062.3	316.4	1005.3	1791.1	892.0	1830.0	1968.8
1993–1994	15359.4	4324.7	496.1	2234.1	340.1	1011.0	1916.9	949.4	2030.7	2056.4
1994–1995	16333.1	4522.0	542.2	2470.9	372.0	1065.4	2114.8	1034.0	2108.5	2103.3
1995–1996	17541.5	4493.1	574.0	2851.7	397.3	1129.1	2428.6	1135.9	2274.5	2257.2
1996–1997	18877.5	4909.3	577.2	3119.0	418.9	1150.3	2626.1	1226.8	2411.6	2438.2
1997–1998	19741.2	4802.7	633.8	3129.3	451.3	1270.7	2827.4	1303.5	2682.0	2640.3
1998–1999	21014.8	5084.5	651.7	3226.2	483.0	1350.5	3047.0	1386.9	2888.5	2896.5
1999–2000	22549.4	5227.9	679.0	3384.6	508.9	1463.8	3268.9	1507.2	3271.1	3238.0
2000–2001	23484.8	5227.5	694.7	3631.6	520.3	1553.8	3437.5	1645.5	3386.6	3387.2
2001–2002	24749.6	5541.6	707.7	3714.1	529.9	1615.7	3768.9	1752.3	3596.8	3522.7
2002–2003	25709.4	5175.6	767.2	3969.1	555.0	1749.6	4027.6	1951.4	3856.6	3657.2
2003–2004	27757.5	5643.9	787.9	4220.6	580.6	1967.1	4433.4	2213.0	4061.0	3850.0
2004–2005	29714.7	5654.3	850.3	4532.2	626.8	2288.6	4773.0	2504.2	4371.7	4113.6

(continued)

Table A 6.1 (continued)

Year	Total GDP	Agriculture, forestry and fishing	Mining and querying	Manufacturing	Electricity, gas and water supply	Construction	Trade, hostels and restaurants	Transport, storage and communi-cation	Financing, insurance, real estate and business services	Community, social and personal services
2005–2006	32530.7	5944.9	861.4	4990.2	671.2	2581.3	5354.0	2800.1	4923.4	4404.3
2006–2007	35643.6	6191.9	925.8	5704.6	733.6	2848.1	5949.2	3151.7	5610.6	4528.2
2007–2008	38966.4	6550.8	960.0	6290.7	794.3	3155.0	6550.1	3545.1	6281.2	4839.2
2008–2009	41586.8	6556.9	980.6	6563.0	830.5	3323.3	6922.2	3929.0	7036.3	5445.0
2009–2010	45160.7	6609.9	1038.3	7304.4	882.2	3544.4	7470.2	4508.7	7719.1	6083.7
2010–2011	49185.3	7178.1	1106.2	7951.5	928.6	3747.1	8363.5	5076.7	8491.9	6341.7
2011–2012	52475.3	7538.3	1107.3	8541.0	1006.5	4151.9	8466.3	5556.3	9455.3	6652.5

References

Ahluwalia IJ (1995) New economic policies, enterprises and privatisation in India. In Cassen R, Joshi V (eds) India: future of economic reforms. Oxford University Press, New Delhi

Bardhan P (1984) The political economy of development in India. Oxford University Press, New Delhi

Dandekar VM (1992) Forty years after Independence. In: Jalan B (ed) The Indian economy: problems and prospects. Penguin India, New Delhi

Dhar PN (1988) The Indian economy: past performance and current issues. In: Lucas REB, Papanek GF (eds) The Indian economy: recent development and future prospects. Oxford University Press, New Delhi

Government of India (2008) National accounts statistics: manual on estimation of state and district income 2008. Ministry of Statistics and Programme Implementation. Central Statistics Office, New Delhi

Government of India (2014) Statistical year book India 2014. Ministry of Statistics and Programme Implementation, New Delhi

Kitagawa EM (1955) Components of a difference between two rates. J Am Stat Assoc 50(272):1168–1194

Meinen G, Verbiest P, de Wolf P-P (1998) Perpetual inventory method. Service lives, discard patterns and depreciation methods. Department of National Accounts, Voorburg, Statistics Netherlands

Nagraj R (1990) Growth rate of India's GDP, 1950-51 to 1987-88: examination of alternative hypotheses. Econ Polit Weekly 25(6):13961403

Nicholson JL (1955) National income at factor cost or market prices? Econ J 65(258):216–224

OECD/Eurostat (2012) Component expenditures of GDP. In: Eurostat-OECD Methodological Manual on Purchasing Power Parities. Paris OECD

Patnaik P (1987) Recent growth experience of the Indian economy. Econ Polit Weekly, May 1990

Raj KN (1984) Some observations on economic growth in India over the period 1952-53 to 1982-83. Econ Polit Weekly 19(41)

Ryan SE, Porth LS (2007) A tutorial on the piecewise regression approach applied to bed-load transport data. Fort Collins, US Department of Agriculture, Forest Service, Rocky Mountain Research Station. Gen. Tech. Rep. RMRS-GTR-189

Virmani A (1989) Growth, capital and employment. Performance in perspective. Economic Research Working Paper No. 3. Planning Commission, New Delhi

Virmaini (1997a) India: crises, reform and growth. Econ Polit Weekly, August 9

Virmani (1997b) Economic development and transition: India. Paper presented at the Tokyo dialogue on alternatives to the World Bank–IMF approach to reforms and growth. Economic Planning Agency, Tokyo, 7 December 1997

Virmani (2005) India's economic growth history: fluctuations, trends, break points and phases. International Council for Research on International Economic Relations. Working Paper, New Delhi

Chapter 7
Demographic Dividend, 2001–2011

Abstract This chapter quantifies the contribution of demographic transition to the economic growth in India and in its constituent states/Union Territories during 2001–2011 in terms of the transition in the child support ratio, ageing index and the proportion of the old population as the three determine the ratio of the working age population to the total population. The analysis suggests that the demographic dividend resulting from age structure transition contributed only marginally to economic growth that India and its constituent states/Union Territories have witnessed during 2001–2011. Moreover, demographic dividend varied widely across states/Union Territories and reversed in one state.

7.1 Introduction

It is well established that demographic transition has important implications for economic growth through changes in the age structure of the population. Bloom and Williamson (1998) were the first to demonstrate that demographic transition played a key role in the economic miracle in selected East Asian countries in the 1990s. Subsequently, many studies have been carried out in different parts of the world to highlight the contribution of demographic change to economic growth (Bloom et al. 2001; Bloom et al. 2003; Bloom and Finlay 2008; Bloom 2011; Golley and Tyres 2011; Joe et al. 2015; Mason et al. 2008; Prskawetz et al. 2007; Ranganathan et al. 2015; Wang et al. 2013). The argument is that population age structure can have significant impact on the performance of the economy. In populations with high proportion of children, a high proportion of resources are needed to meet the essential needs of children which tend to depress the pace of economic growth. On the other hand, if most of the population falls within working ages, the added productivity of this group can produce a 'demographic dividend' that spurs economic growth, assuming that policies to take advantage of this dividend are in place. The effect of a large working age population combined with health, family, labour, financial and human capital-related policies, in fact, can lead to a virtuous cycle of wealth creation. Finally, if a large proportion of the population comprises the old population, a large

© Springer Nature Singapore Pte Ltd. 2020

A. R. Chaurasia, *Population and Sustainable Development in India*,
https://doi.org/10.1007/978-981-32-9212-3_7

share of resources is again needed by the relatively less productive segment of the population inhibiting economic growth (Bloom et al. 2003).

Mason (2005) and Mason and Lee (2006) have differentiated between the first and the second demographic dividend. The first demographic dividend is defined in terms of the conventional child support ratio. It is positive when the working age population raises relatively fewer number of children as a result of decrease in fertility while the number of old age dependents is still low because the life expectancy remains low. During this period, the working age population grows more rapidly than the population dependent on it, resulting in the availability of more resources for investment in economic development and family welfare. As a result, per capita income also increases more rapidly. The period of this dividend may last for five decades or even more, but at later stages of demographic transition, the first demographic dividend turns negative as the working age population decreases while the old age population increases rapidly so that the potential support ratio or the ratio of the working age population to the old population decreases.

The potential support ratio, however, is not the only factor that affects the growth of per capita income. People, in general, have the tendency of creating assets and accumulating wealth as they get older. People also rely more on returns on their assets and wealth than on the work as they get older. An increase in assets created and an increase in the individual wealth lead to increased investments, thereby increasing the national income and hence per capita income. Recognising this fact, Lee et al. (2000) introduced the concept of the second demographic dividend which operates through productivity growth by inducing the accumulation of wealth and hence capital. The first demographic dividend yields a bonus which is essentially transitory in nature. The second demographic dividend transforms the transitory bonus resulting from the first demographic dividend into greater assets and sustainable development. The second dividend depends upon the first and begins somewhat later but, unlike the first dividend which occurs for a specific duration, it can continue indefinitely. The two dividends, of course, are not automatic. They depend upon the implementation of effective policies and, therefore, provide, at best, an opportunity rather than a guarantee of improved standards of living.

In recent years, there has been considerable interest in the role of demographic transition in economic growth in India. The demographic dividend resulting from the change in population age structure as a result of decrease in fertility and mortality has been pointed out as the basis for optimism for India's economic future (Bloom 2011; Kelkar 2004; Kumar 2010; Aiyar and Mody 2011). This optimism has been shared at the official level also (Government of India 2013). However, Kumar and Subramanian (2012) have observed that the so-called demographic dividend in India was strong and positive during the 1990s but, during the 2000s, either there has been no effect of the change in the population age structure on economic growth or the effect has been negative. Similarly, Thakur (2012) has also concluded that the growth in the working age ratio has been negatively influencing the economic growth. Chandrashekhar et al. (2006) have also concluded that India has not been able to exploit the benefit of the change in population age structure because of the unsatisfactory employment, education and health situation. Recently, Singh (2016) has also observed that under the conditions prevailing in the country, the high optimism about India's ability to reap the demographic dividend seems to be misplaced.

Nearly all studies related to India's demographic dividend are based on data till 2001. There are only two studies that cover the post 2001 period (Kumar and Subramanian 2012; Thakur 2012). Both these studies have, however, been based on the projected not the actual population growth in the country during 2001–2011. It is well known that the actual population enumerated at the 2011 population census was larger than the population projected by the Government of India on the basis of results of the 2001 population census. Moreover, both the studies have focused on the growth of the working age population relative to the growth of the total population. It may, however, be pointed out that the growth of the working age population is contingent upon the growth of the child population and the growth of the old population. The growth of the child population depends primarily upon the transition in fertility, whereas the growth of the old population depends, largely, upon the chance of surviving to old ages or upon mortality transition (United Nations 2002; Grundy 1996). This means that any analysis of the effect of the demographic transition on the economic growth should be carried out in terms of the change in the proportion of the child population resulting primarily from fertility transition and the change in the proportion of the old population resulting from mortality transition. Such an analysis is necessary to understand why demographic transition has contributed only marginally to economic growth in India during 2001–2011, as observed by Kumar and Subramanian (2012) and Thakur (2012).

The above considerations constitute the rationale for the present chapter which uses the data available through India's 2001 and 2011 population census to quantify the contribution of demographic transition to the economic growth in India and in its constituent states/Union Territories during 2001–2011. The chapter analyses demographic transition in terms of the transition in the child support ratio, ageing index and the proportion of the old population as the three factors that determine the ratio of the working age population to the total population. The analysis has been carried out for all except three states/Union Territories of the country for which estimates of gross domestic products (GDP) are prepared by the Government of India. Estimates of GDP are not available for the Union Territories of Daman and Diu, Dadra and Nagar Haveli, and Lakshadweep. Moreover, the analysis has been carried out in both relative (growth of per capita GDP) and absolute (increase in per capita GDP) terms.

The rest of the chapter is organised as follows: The next section describes the methodology used for analysing the contribution of the change in the age composition of the population to the change in per capita GDP following the factor decomposition approach. Section three describes the data source. Section four analyses per capita GDP growth in India and its states/Union Territories during the period 2001–2011. Section five analyses the absolute increase in per capita GDP in real terms and measures the relative contribution of change in demographic and non-demographic factors to the change in per capita GDP in the country and in its selected states/Union Territories. Section six discusses the findings of the analysis in the context of demographic transition. The last section highlights the demographic imperatives for India in the context of economic growth.

7.2 Methodology

Let Y denotes the gross domestic product (GDP) at constant prices, L denotes the number of workers, W denotes the working age population and P denotes the population. Then, it is well known that the per capita real GDP may be written as:

$$\frac{Y}{P} = \frac{Y}{L} * \frac{L}{W} * \frac{W}{P} \tag{7.1}$$

or

$$i = p * u * d \tag{7.2}$$

Here i is the per capita real GDP, p is the average productivity of workers, u is the extent of the utilisation of the working age population or the ratio of the workers to the working age population and d is the ratio of the working age population to the total population. Equation (7.2) suggests that the level of per capita real GDP depends upon three factors—average productivity of the worker, productive utilisation of the working age population and a demographic component. The change in the per capita real GDP, therefore, is the result of the change in the average productivity of the worker, change in the productive utilisation of the working age population and change in the demographic component. This change can be captured in relative as well as absolute terms. In relative terms, the change in per capita real GDP can be decomposed as:

$$r^i = r^p + r^u + r^d \tag{7.3}$$

where

$$r^i = ln\left(\frac{r_2^i}{r_1^i}\right) \tag{7.4}$$

The relative change in the demographic component, r^d, may be decomposed as:

$$r^d = ln\left(\frac{W_2/P_2}{W_1/P_1}\right) = ln\left(\frac{W_2}{W_1}\right) - ln\left(\frac{W_2}{W_1}\right) = r^w - r^n \tag{7.5}$$

Equation (7.5) suggests that r^d is positive when $r^w > r^n$. In this case, the change in the demographic component contributes to accelerating the increase in r^i even if there is no increase in r^p and r^u. However, when $r^w < r^n$, r^d is negative so that the change in the demographic component contributes to decelerating the increase in per capita GDP. Finally, when $r^w = r^n$, $r^d = 0$ so that the change in the demographic component neither accelerates nor decelerates the increase in r^i.

One limitation of Eq. (7.3) is that it treats the three components of the change in per capita real GDP independent of each other. It may also be noticed that Eq. (7.3) is true by definition so that the naive regression or correlation approaches, which ignore the sum constraint, are potentially problematic (Poorter and van der Werf 1998; Wright and Westoby 2001). A more pragmatic approach is to decompose the variance in the per capita GDP into the variance in its three components (Preston 1994). Notice that the variance in the growth in real GDP per capita may be written as:

$$Var\left(r^i\right) = Var(r^p + r^u + r^d) \tag{7.6}$$

or using only suffixes,

$$\begin{aligned} Var(i) = {} & [var(p) + cov(p, u) + cov(p, d)]+ \\ & [var(u) + cov(u, p) + cov(u, d)]+ \\ & [var(d) + cov(d, p) + cov(d, u) \end{aligned} \tag{7.7}$$

where *var* denotes the variance and *cov* denotes the covariance. The first term of the right side of Eq. (7.7) gives the contribution of the variation in the growth of worker productivity to the variation in the growth of per capita real GDP. Similarly, the second term gives the contribution of the variation in increase in the utilisation of the working age population while the third term gives the contribution of the variation in the change in the demographic component to the variation in growth in per capita real GDP. Dividing both sides of Eq. (7.7) by *Var*(i), the proportionate contribution of the variance in r^p to the variance in r^i is given by:

$$C_p = \frac{Var(p) + cov(p, u) + cov(p, d)}{var(i)} \tag{7.8}$$

It may, however, be noticed that the relative contribution of the variation in one of the components of the growth in per capita real GDP to the variation in the growth in per capita real GDP may be small for two counts. First, the component is not variable or the variation is very small and so the variance and covariance terms in Eq. (7.7) are small. Second, the component is variable but the covariance terms are negative so that the algebraic sum of the variance and the covariance terms is small. In the second case, Eq. (7.7) does not reflect the true contribution of the variation in a component to the variation in the growth in per capita real GDP simply because some of the covariance terms are negative.

The above problem may be circumvented and the true importance of the variance in different components to the variance in the growth of per capita real GDP may be assessed by using absolute instead of actual values of covariance terms. For example, the importance of the variance in the growth of worker productivity (r^p) to the variance in the growth of per capita real GDP (r^i) may be obtained as:

$$I_p = \frac{Var(p) + |cov(p, u)| + |cov(p, d)|}{var(i)} \tag{7.9}$$

where the normalising constant T is the sum of the absolute values of the terms on the right-hand side of Eq. (7.7). Similar approach has been used by Horvitz et al. (1997) and Rees et al. (1996, 2010) in other contexts. This approach does not attempt to assign the causality but it explicitly shows the relative importance of the variation in one of the components of per capita GDP growth to the variation in per capita GDP growth.

The methodological considerations discussed above are confined to the relative change or the growth in per capita real GDP and its decomposition in terms of the increase in worker productivity, utilisation of the working age population and change in the demographic component. Following Kim and Strobino (1984), the absolute change in the per capita GDP between two points in time can be decomposed as:

$$\nabla i = i_2 - i_1 = (p_2 * u_2 * d_2) - (p_1 * u_1 * d_1)$$

or

$$\nabla i = (p_2 - p_1) * u_1 * d_1 + p_1 * (u_2 - u_1) * d_1 + p_1 * u_1 * (d_2 - d_1) +$$

$$(p_1 - p_2) * (u_2 - u_1) * d_1 + (p_1 - p_1) * u_1 * (d_2 - d_1) +$$

$$p_1 * (u_2 - u_1) * (d_2 - d_1) + (p_2 - p_1) * (u_2 - u_1) * (d_2 - d_1) \tag{7.10}$$

Equation (7.10) shows how interaction effects contribute to the absolute change in the per capita real GDP. For example, the contribution of the change in the demographic component to the change in per capita real GDP may be direct as well as indirect working through the change in the worker productivity and the change in the productive utilisation of the working age population.

7.3 Data Source

Data from two sources have been used in the present analysis. Estimates of GDP for India and its constituent states and Union Territories are the official estimates prepared by the Government of India for the year 2001–2002 and 2011–2012 at

2004–2005 prices to exclude the effect of inflation. These estimates are available for all states and Union Territories except three Union Territories—Dadra and Nagar Haveli, Daman and Diu, and Lakshadweep. As such, the present analysis excludes the three Union Territories. On the other hand, estimates of total population, child population, working age population, old population and total workers have been derived from 2001 and 2011 population census. The child population is the population below 15 years of age; working age population, in the present analysis, is taken as the population aged 15–59 years as the population census does not provide estimates of the number of workers aged 15–64 years. Finally, the old population is the population aged at least 60 years. In India's population census, a person is classified as worker if the person concerned has worked even for a day during the year prior to the census. A comprehensive definition of work including the type of work and the duration of work was adopted at the population census to classify a person as a worker (Government of India 2001). Moreover, workers aged 15–59 years have been used in the present analysis. Table 7.1 presents the basic data used in the present analysis for the country as a whole and for its constituent states and Union Territories.

7.4 GDP Growth 2001–2011

India's GDP at 2004–2005 prices is estimated to have more than doubled, from Rs 22,142 billion during 2001–2002 to Rs 46,826 billion during 2011–2012, at an average annual growth rate of 7.49% per year. The growth of the economy was relatively most rapid in Sikkim where the real GDP increased at an average annual rate of more than 13% per year, followed by Uttarakhand. Sikkim and Uttarakhand are the only two states/Union Territories in the country where the real GDP recorded double-digit growth on average during the 10 years between 2001–2002 and 2011–2012. In contrast, the growth of the economy was relatively the slowest in Assam where the real GDP increased at the rate of just around 5% per year on average. In addition to Assam, the growth of real GDP was estimated to be less than 6% per year on average in Jammu and Kashmir, Manipur, Punjab and Uttar Pradesh during the period under reference. There are 15 states/Union Territories where the growth of the real GDP during the period under reference was slower than the national average, whereas in 17 states/Union Territories the growth rate of real GDP was estimated to be faster than the national average (Fig. 7.1). In other words, the inter-state/Union Territories inequality in the growth of the real GDP in the country during 2001–2012 was quite substantial as the Theil index of inequality in real GDP growth across the states/Union Territories is estimated to be 0.065. One component of this inequality is the difference in the growth of population across the states/Union Territories of the country.

Table 7.1 Real GDP and population in India and states/Union Territories, 2001 and 2011

India/State/Union Territory	Real GDP		Population		Population (15–59)		Workers (15–59)	
	Billion rupees		Million		Million		Million	
	2001	2011	2001	2011	2001	2011	2001	2011
India	22141.270	46825.977	1025.433	1205.715	585.000	729.631	358.000	426.668
AN Islands	14.048	34.596	0.354	0.380	0.232	0.262	0.128	0.144
Andhra Pradesh	1774.887	3824.589	76.077	83.809	45.890	53.740	31.140	34.898
Arunachal Pradesh	24.379	52.204	1.097	1.382	0.604	0.825	0.434	0.532
Assam	441.973	734.444	26.627	31.187	15.096	18.859	8.524	10.776
Bihar	686.202	1301.709	82.821	103.696	42.445	54.268	24.264	29.592
Chandigarh	57.204	133.378	0.899	1.055	0.593	0.722	0.327	0.386
Chhattisgarh	344.125	789.030	20.806	25.522	11.609	15.334	8.585	10.952
Delhi	760.596	1807.654	13.829	16.771	8.617	11.058	4.341	5.281
Goa	95.912	224.991	1.341	1.456	0.897	0.975	0.488	0.531
Gujarat	1388.254	3675.813	50.624	60.201	30.501	37.969	19.604	22.512
Haryana	700.270	1637.702	21.064	25.320	11.900	15.596	7.577	8.107
Himachal Pradesh	187.359	390.544	6.063	6.854	3.631	4.376	2.605	3.077
Jammu and Kashmir	230.161	382.700	10.098	12.526	5.806	7.363	3.274	3.874
Jharkhand	438.456	894.911	26.913	32.871	14.625	18.624	9.039	11.589
Karnataka	1365.159	2727.213	52.799	61.049	31.891	39.234	21.110	25.075
Kerala	904.501	1898.507	31.815	33.372	20.183	21.347	9.489	10.541
Madhya Pradesh	955.251	1781.435	60.189	72.544	32.655	42.528	22.716	28.102
Maharashtra	3184.386	7420.418	96.761	111.967	57.206	70.942	36.947	43.865
Manipur	39.714	66.806	2.162	2.848	1.310	1.786	0.833	1.156
Meghalaya	51.691	104.131	2.316	2.961	1.229	1.644	0.850	1.060
Mizoram	21.223	49.793	0.888	1.097	0.525	0.672	0.411	0.443

(continued)

Table 7.1 (continued)

India/State/Union Territory	Real GDP		Population		Population (15–59)		Workers (15–59)	
	Billion rupees		Million		Million		Million	
	2001	2011	2001	2011	2001	2011	2001	2011
Nagaland	43.585	92.540	1.985	1.977	1.167	1.195	0.737	0.837
Odisha	564.750	1251.311	36.743	41.855	21.496	25.794	12.677	15.646
Puducherry	52.298	108.064	0.974	1.247	0.630	0.828	0.321	0.417
Punjab	829.809	1476.701	24.265	27.702	14.456	17.751	8.156	8.763
Rajasthan	1012.634	2130.793	56.220	68.279	29.866	39.441	20.899	26.664
Sikkim	12.920	47.843	0.539	0.609	0.321	0.403	0.230	0.276
Tamil Nadu	1851.012	4034.157	61.977	72.080	39.758	47.563	24.923	29.298
Tripura	64.057	143.867	3.194	3.672	1.886	2.365	1.043	1.345
Uttar Pradesh	2218.432	3963.094	165.461	198.190	85.888	111.442	46.430	55.793
Uttarakhand	175.713	556.668	8.472	10.070	4.730	6.040	2.774	3.392
West Bengal	1650.307	3088.371	80.064	91.164	47.719	58.684	26.765	31.744

Source Government of India

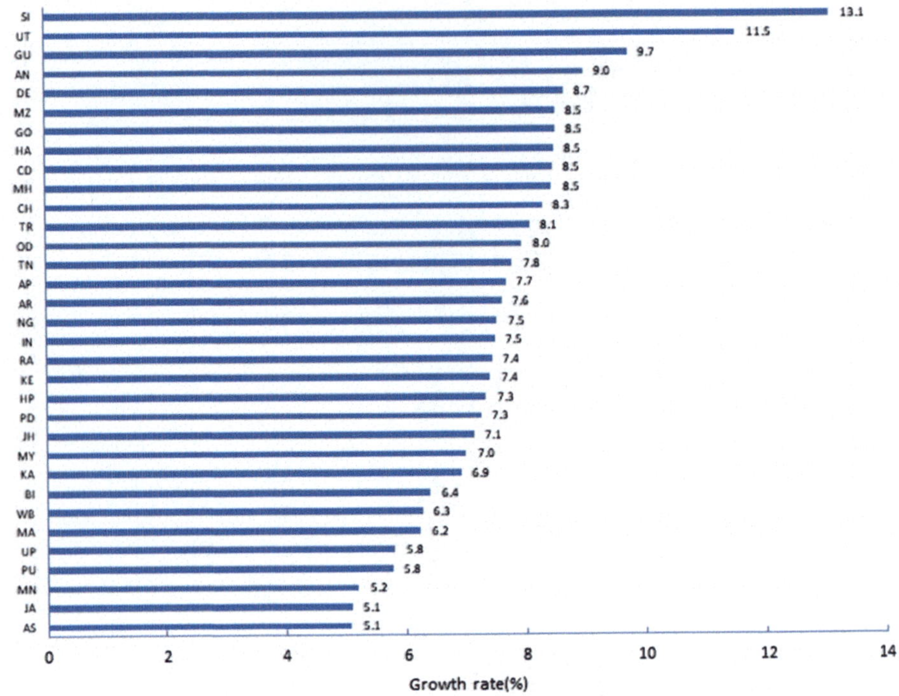

Fig. 7.1 Growth rate of real GDP in India and states/Union Territories, 2011–2012

7.5 Population Growth

According to the data available through 2001 and 2011 population census, the population of the country increased from 1025 million in 2001 to 1206 million in 2011 at an average annual growth rate of around 1.62% per year. This means that more than 180 million people were added to the population of the country during the 10 years between 2001 and 2011. The growth of the population was the fastest in Manipur where the population increased at an average annual rate of almost 2.76% per year during the period under reference. This means that population of Manipur was at an early stage of transition at the 2011 population census. In addition, there are nine states/Union Territories where population increased at an average annual rate of at least 2% per year during 2001–2011 implying slow population transition. In contrast, Nagaland is the only state where the population decreased, instead increased, during the period under reference. In addition, there are only three states and Union Territories—Andaman and Nicobar Islands, Andhra Pradesh and Kerala—where the population increased at an average annual rate of less than 1% per year with Kerala recording the slowest growth of population in the country at an average annual growth rate of less than 0.5% per year, which implies that the population of Kerala was at an advanced stage of transition at the time of the 2011 population census. Similarly,

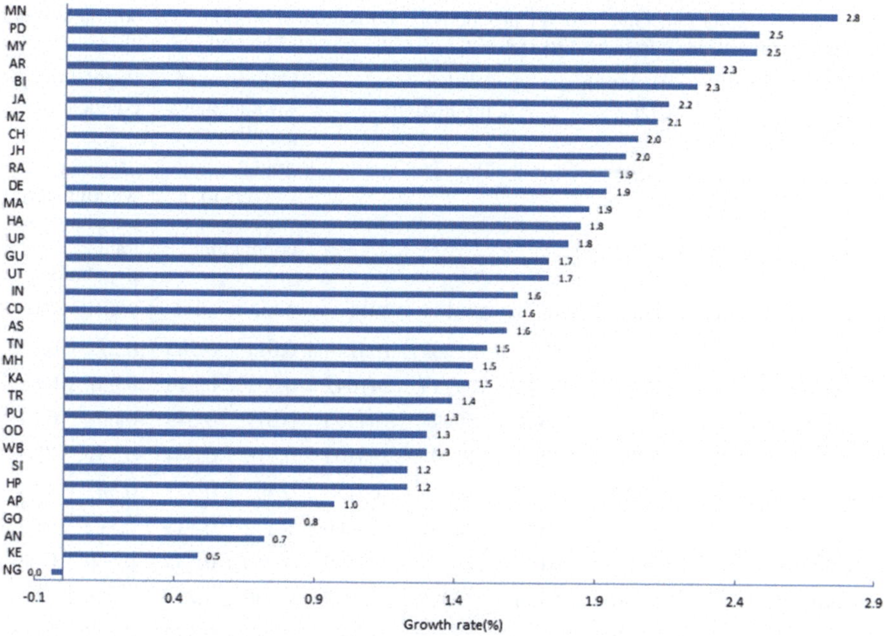

Fig. 7.2 Population growth rate in India and states/Union Territories, 2001–2011

population transition in Andaman and Nicobar Islands and Andhra Pradesh appears to be quite advanced (Fig. 7.2 and Table 7.2).

7.6 Growth of Per Capita GDP

As a result of the growth in real GDP and increase in the population, the per capita real GDP of the country increased by almost 80%, from around Rs 21,592 in 2001–2002 to around Rs 38,837 in 2011–2012, at an average annual growth rate of around 5.87% per year. The growth rate of per capita real GDP was slower than the growth rate of real GDP because the population of the country also increased at around 1.62% per year during this period. Had the population decreased, instead increased, during this period, the growth of per capita real GDP would have been faster than the growth of real GDP (Fig. 7.3).

Among different states/Union Territories of the country, the growth of per capita real GDP was slower than the growth of real GDP in all but one state/Union Territory (Table 7.3). The only state where the growth of per capita real GDP was faster than that of the real GDP is Nagaland because the population of the state decreased, instead increased, during the 10 years between 2001 and 2011. It is also obvious from Table 7.3 that the faster the increase in the population, the slower the growth of the real per capita GDP in relation to the growth of the real GDP.

Table 7.2 Per capita real GDP, worker productivity, utilisation of working age population and population composition in India and its states

India/State/Union territory	Per capita real GDP		Worker productivity		Utilisation of working age population		Population composition	
	Rupees		Rupees		Ratio of workers to working age population		Ratio of working age population to total population	
	2001	2011	2001	2011	2001	2011	2001	2011
India	21592	38837	61909	109748	0.611	0.585	0.571	0.605
AN Islands	39699	90992	109749	241042	0.551	0.548	0.657	0.689
Andhra Pradesh	23330	45634	56997	109593	0.679	0.649	0.603	0.641
Arunachal Pradesh	22226	37768	56201	98204	0.718	0.644	0.551	0.597
Assam	16599	23550	51849	68153	0.565	0.571	0.567	0.605
Bihar	8285	12553	28281	43989	0.572	0.545	0.512	0.523
Chandigarh	63607	126412	175199	345522	0.550	0.535	0.660	0.684
Chhattisgarh	16540	30916	40086	72042	0.740	0.714	0.558	0.601
Delhi	54999	107783	175223	342266	0.504	0.478	0.623	0.659
Goa	71541	154479	196501	423931	0.544	0.544	0.669	0.669
Gujarat	27423	61059	70815	163284	0.643	0.593	0.602	0.631
Haryana	33245	64680	92415	202014	0.637	0.520	0.565	0.616
Himachal Pradesh	30901	56977	71920	126929	0.717	0.703	0.599	0.638
Jammu and Kashmir	22793	30552	70293	98788	0.564	0.526	0.575	0.588
Jharkhand	16292	27225	48508	77218	0.618	0.622	0.543	0.567
Karnataka	25856	44672	64668	108760	0.662	0.639	0.604	0.643
Kerala	28430	56890	95317	180107	0.470	0.494	0.634	0.640
Madhya Pradesh	15871	24557	42052	63393	0.696	0.661	0.543	0.586
Maharashtra	32910	66273	86188	169167	0.646	0.618	0.591	0.634
Manipur	18367	23457	47680	57786	0.636	0.647	0.606	0.627
Meghalaya	22322	35166	60784	98260	0.692	0.645	0.531	0.555
Mizoram	23905	45408	51579	112316	0.784	0.660	0.591	0.613
Nagaland	21954	46808	59164	110525	0.631	0.700	0.588	0.605
Odisha	15370	29896	44550	79976	0.590	0.607	0.585	0.616
Puducherry	53720	86669	162925	259404	0.510	0.503	0.647	0.664
Punjab	34198	53307	101746	168506	0.564	0.494	0.596	0.641
Rajasthan	18012	31207	48455	79913	0.700	0.676	0.531	0.578
Sikkim	23975	78508	56270	173550	0.715	0.685	0.596	0.661
Tamil Nadu	29866	55968	74269	137694	0.627	0.616	0.642	0.660
Tripura	20053	39176	61391	106928	0.553	0.569	0.591	0.644

(continued)

Table 7.2 (continued)

India/State/Union territory	Per capita real GDP		Worker productivity		Utilisation of working age population		Population composition	
	Rupees		Rupees		Ratio of workers to working age population		Ratio of working age population to total population	
	2001	2011	2001	2011	2001	2011	2001	2011
Uttar Pradesh	13408	19996	47780	71033	0.541	0.501	0.519	0.562
Uttarakhand	20741	55282	63348	164094	0.586	0.562	0.558	0.600
West Bengal	20612	33877	61658	97291	0.561	0.541	0.596	0.644

Source Author's calculations based on Table 7.1

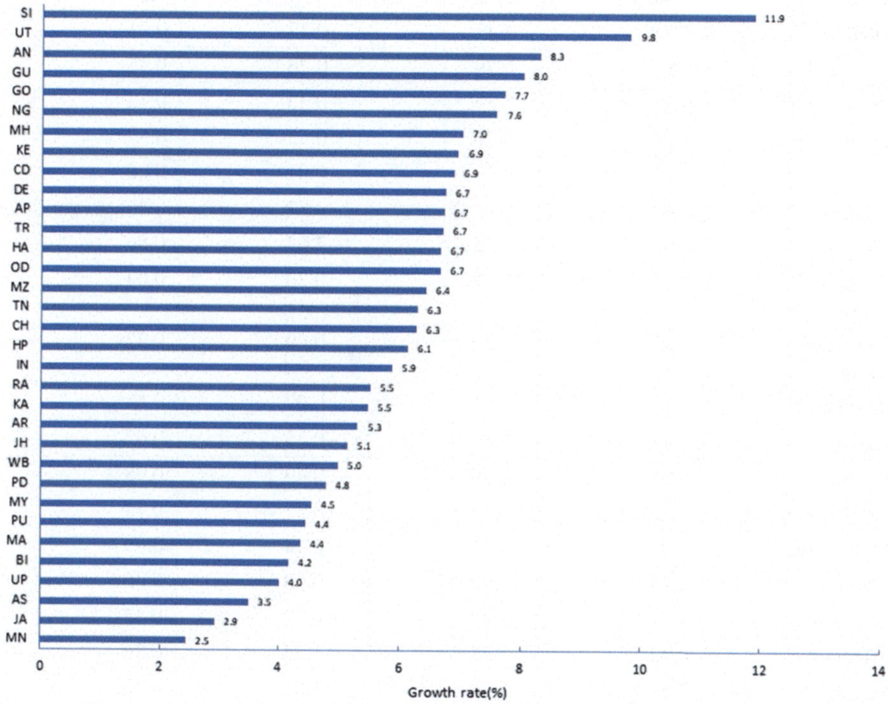

Fig. 7.3 Growth rate of per capita real GDP in India and its states/Union Territories

Table 7.3 Average annual change (%) in per capita real GDP, worker productivity, productive utilisation of working age population and ratio of workers to working age population during 2001–2011 in India, and its states and Union Territories

India/State/Union Territory	Average annual change (%) in			
	Per capita real GDP	Worker productivity	Productive utilisation of working age population	Ratio of working age population to total population
	r^i	r^p	r^u	r^d
India	5.87	5.73	−0.44	0.58
AN Islands	8.29	7.87	−0.06	0.48
Andhra Pradesh	6.71	6.54	−0.44	0.61
Arunachal Pradesh	5.30	5.58	−1.09	0.81
Assam	3.50	2.73	0.12	0.65
Bihar	4.15	4.42	−0.47	0.21
Chandigarh	6.87	6.79	−0.28	0.36
Chhattisgarh	6.25	5.86	−0.35	0.74
Delhi	6.73	6.70	−0.53	0.57
Goa	7.70	7.69	0.01	0.00
Gujarat	8.00	8.35	−0.81	0.46
Haryana	6.66	7.82	−2.03	0.86
Himachal Pradesh	6.12	5.68	−0.20	0.64
Jammu and Kashmir	2.93	3.40	−0.69	0.22
Jharkhand	5.13	4.65	0.07	0.42
Karnataka	5.47	5.20	−0.35	0.62
Kerala	6.94	6.36	0.49	0.08
Madhya Pradesh	4.36	4.10	−0.51	0.77
Maharashtra	7.00	6.74	−0.44	0.69
Manipur	2.45	1.92	0.18	0.35
Meghalaya	4.54	4.80	−0.71	0.45
Mizoram	6.42	7.78	−1.72	0.36
Nagaland	7.57	6.25	1.04	0.28
Odisha	6.65	5.85	0.28	0.52
Puducherry	4.78	4.65	−0.13	0.26
Punjab	4.44	5.04	−1.33	0.73
Rajasthan	5.50	5.00	−0.34	0.84
Sikkim	11.86	11.26	−0.44	1.04

(continued)

Table 7.3 (continued)

India/State/Union Territory	Average annual change (%) in			
	Per capita real GDP	Worker productivity	Productive utilisation of working age population	Ratio of working age population to total population
	r^i	r^p	r^u	r^d
Tamil Nadu	6.28	6.17	−0.17	0.28
Tripura	6.70	5.55	0.28	0.87
Uttar Pradesh	4.00	3.97	−0.77	0.80
Uttarakhand	9.80	9.52	−0.43	0.72
West Bengal	4.97	4.56	−0.36	0.77

Source Author's calculations

On the other hand, the increase in the per capita real GDP or the per capita real income during 2001–2011 was relatively the fastest in Sikkim which is the only state/Union Territory in the country where the per capita real GDP increased at an average annual rate of increase of more than 10% per year (Table 7.3). In addition, there are six states/Union Territories—Andaman and Nicobar Islands, Goa, Gujarat, Maharashtra, Nagaland and Uttarakhand—where the per capita real GDP increased at an average annual rate of more than 7% per year but less than 10% per year. By comparison, the increase in the per capita real GDP was the slowest in Manipur where the per capita real GDP increased at an average annual rate of less than 2.5% per year. In Jammu and Kashmir also, per capita real GDP increased at an average annual rate of increase of less than 3% per year during the 10 years between 2001–2002 and 2011–2012. In 18 states/Union Territories of the country, however, the increase in the per capita real GDP was faster than the national average during the period under reference.

7.7 The Demographic Dividend

The change in the per capita real GDP over time is the algebraic sum of the change in the productivity of the worker, change in the productive utilisation of the working age population and the change in the ratio of the working age population to the total population. In India, the per capita real GDP increased at an average annual rate of 5.87% per year during 2001–2012. During the same period, worker productivity in the country increased at an average annual rate of almost 5.73% per year while the productive utilisation of the working age population recorded a decrease at an average annual rate of around 0.44% per year. Finally, the ratio of the working age population to the total population increased at an average annual rate of around 0.58% per year. The increase in the ratio of the working age population to the total population confers a demographic dividend in the rate of increase in per capita real GDP. Table 7.3 indicates that the rate of increase in the per capita real GDP attributed to the demographic dividend constituted only a small proportion of the rate of increase in the per capita real GDP in the country during 2001–2012. Almost 98% of the increase in the per capita real GDP in the country during 2001–2012 was attributed to the increase in the worker productivity whereas the decrease in the productive utilisation of the working age population decreased during the period under reference inducing a deceleration, not acceleration, in the rate of increase in the per capita real GDP.

Among different states/Union Territories, the increase in the ratio of the working age population to the total population was the highest in Sikkim (1.04% per year) but the lowest in Goa where there was virtually no change in this ratio. In Kerala also, this ratio increased only marginally. By comparison, this ratio increased very rapidly—at least 0.80% per year on average—in Arunachal Pradesh, Haryana, Rajasthan, Tripura and Uttar Pradesh, whereas it increased very slowly—less than 0.30% per year—in Bihar, Jammu and Kashmir, Nagaland, Puducherry and Tamil Nadu.

The demographic dividend during the period under reference was, however, relatively the highest in Uttar Pradesh where the increase in the ratio of the working age population to the total population accounted for more than 20% of the increase in the per capita real GDP, although the average annual rate of increase in this ratio was not the highest in the country. There are only 13 states/Union Territories in the country, where the increase in the ratio of the working age population to the total population or the demographic dividend accounted for more than 10% of the increase in the per capita real GDP during 2001–2012. On the other hand, the demographic dividend accounted for virtually no increase in per capita real GDP in Goa whereas in Kerala, Nagaland and Tamil Nadu, the demographic dividend accounted for less than 5% of the increase in the per capita real GDP. There are 19 states/Union Territories where the demographic dividend in the country resulting from the increase in the ratio of the working age population to the total population was smaller than the national average. It is, however, clear from Table 7.3 that the demographic dividend in the country was never large during the period under reference.

In absolute terms, the per capita real GDP in India increased by Rs 17,245 between 2001–2002 and 2011–2012 as a result of change in worker productivity, productive utilisation of the working age population and demographic dividend. Assuming that interaction effects are distributed equally among the participating variables, the absolute increase in the per capita real GDP attributed to demographic dividend was Rs 1757 or around 10% which again confirms that the contribution of demographic dividend to India's economic growth during 2001–2012 has been quite small. Moreover, only around 74% of the contribution of the demographic dividend was direct; another around 29% was indirect through the increase in worker productivity. On the other hand, demographic dividend had a small negative effect on economic growth through the decrease in the productive utilisation of the working age population and the increase in worker productivity.

Among different states/Union Territories, the direct effect of the demographic dividend on economic growth was the lowest in Sikkim—just around 50% of the total effect but highest in Manipur—almost 90% of the total effect. In addition to Sikkim, Uttarakhand is the only other state in the country where the direct effect of the demographic dividend accounted for less than 60% of the total effect of the demographic dividend on economic growth. In contrast, the direct effect accounted for more than 75% of the total effect of demographic dividend on the economic growth in 14 states/Union Territories of the country.

The indirect effect of the demographic dividend on economic growth has primarily been through the change in the worker productivity and was the highest in Sikkim but lowest in Manipur. The indirect effect of the demographic dividend on per capita real GDP through the increase in the worker productivity was positive in all the states/Union Territories because the worker productivity increased in all the states/Union Territories during the period under reference. By comparison, the indirect effect of the demographic dividend, working through the change in the productive utilisation of the working age population, was positive in eight states/Union Territories because the ratio of the workers to the working age population increased in these states. In the remaining states/Union Territories, the indirect effect of the demographic dividend working through the change in the productive utilisation of the working age population was negative because the ratio of workers to the working age population decreased in these states during 2001–2011. However, the indirect effect of the demographic dividend working through the change in the productive utilisation of the working age population, whether positive or negative, has, at best, been marginal compared to the indirect effect of the demographic dividend working through the change in worker productivity.

Table 7.4 decomposes the variance across states/Union Territories in the increase in per capita real GDP into the variance across states/Union Territories in the increase in the worker productivity, change in the productive utilisation of the working age population and the demographic dividend. The variance across states/Union Territories in the growth of the per capita real GDP during 2001–2012 was almost entirely determined by the variance across states/Union Territories in the increase in the productivity of workers. Compared to the variance in the worker productivity, the contribution of the variance across states/Union Territories in the change in the productive utilisation of the working age population and in the demographic dividend to the variance in per capita real GDP was, at best, marginal.

It may be seen from Table 7.4 that some of the covariance terms are negative so that they mask the relative importance of the variation in a component to the variation in the per capita real GDP across the states/Union Territories. When absolute values of the covariance are used in the analysis, the relative importance of the variation in the increase in worker productivity marginally decreases and that of productive utilisation of the working age population and the demographic dividend increases marginally, although the primary determinant of the variation in the growth of the per capita real GDP across states/Union Territories remains the variation in the growth of the worker productivity. It is clear from Table 7.4 that the importance of the demographic dividend in deciding the variation in the growth in per capita real GDP across states/Union Territories was, at best, marginal.

Table 7.4 Decomposition of the increase in per capita real GDP (Rupees) during 2001–2011 in India, and states and Union Territories

Country/State/Union Territory	Increase in per capita real GDP between 2001–2002 and 2011–2012	Increase in per capita GDP attributed to the increase in			Direct and indirect demographic dividend			
		Worker productivity	Productive utilisation of working age population	Demographic dividend	Direct	Indirect through		
						Worker productivity	Utilisation of working age population	Both
India	17245	16814	−1326	1757	1297	501	−28	−14
AN Islands	51293	48529	−363	3127	1963	1174	−5	−4
Andhra Pradesh	22304	21724	−1517	2097	1470	678	−32	−19
Arunachal Pradesh	15542	16406	−3284	2420	1867	698	−96	−48
Assam	6951	5426	237	1288	1106	174	7	1
Bihar	4268	4543	−494	218	175	49	−4	−1
Chandigarh	62805	62086	−2695	3413	2333	1134	−33	−21
Chhattisgarh	14376	13456	−824	1744	1271	507	−22	−12
Delhi	52784	52543	−4354	4595	3204	1527	−83	−53
Goa	82938	82838	83	17	11	6	0	0
Gujarat	33636	35215	−3607	2029	1284	838	−50	−43
Haryana	31436	37368	−10221	4289	3003	1780	−276	−218
Himachal Pradesh	26076	24169	−882	2789	2039	780	−20	−10
Jammu and Kashmir	7759	9029	−1861	592	510	103	−17	−5
Jharkhand	10933	9882	149	902	693	205	2	1
Karnataka	18816	17876	−1237	2176	1654	564	−29	−13
Kerala	28460	26034	2073	353	238	106	6	4

(continued)

Table 7.4 (continued)

Country/State/Union Territory	Increase in per capita real GDP between 2001–2002 and 2011–2012	Increase in per capita GDP attributed to the increase in			Direct and indirect demographic dividend			
		Worker productivity	Productive utilisation of working age population	Demographic dividend	Direct	Indirect through		
						Worker productivity	Utilisation of working age population	Both
Madhya Pradesh	8686	8166	−1040	1560	1278	324	−32	−11
Maharashtra	33364	32113	−2163	3413	2360	1136	−50	−32
Manipur	5090	3997	371	722	646	69	6	1
Meghalaya	12844	13594	−2050	1300	1032	318	−35	−15
Mizoram	21503	26379	−6127	1251	864	509	−68	−54
Nagaland	24854	20395	3495	964	634	275	35	20
Odisha	14526	12729	631	1165	821	326	12	6
Puducherry	32950	32025	−907	1832	1423	421	−9	−4
Punjab	19109	21814	−5908	3203	2586	849	−162	−71
Rajasthan	13195	11995	−846	2046	1574	511	−27	−12
Sikkim	54533	51551	−2247	5229	2627	2737	−57	−79
Tamil Nadu	26101	25644	−751	1208	855	365	−7	−4
Tripura	19123	15773	824	2527	1815	673	26	13
Uttar Pradesh	6589	6542	−1286	1333	1116	272	−41	−13
Uttarakhand	34540	33480	−1637	2697	1539	1224	−32	−34
West Bengal	13265	12165	−987	2086	1650	477	−29	−11

Source Author's calculations

7.8 Decomposing Demographic Dividend

The demographic dividend, essentially, emanates from the change in the age structure of the population as a result of transition in fertility and mortality under the assumption that the population is closed to migration. The age structure of the population is conventionally characterised in terms of the proportion of the child (0–14 years) population; proportion of the working age (15–59 years) population; and the proportion of the old (60 years and above) population. The demographic dividend (*DD*) is the difference between the growth of the working age population and the growth of the total population. In symbols,

$$DD = r^{w} - r^{n} = \left(r^{w} - r^{c}\right) + \left(r^{c} - r^{o}\right) + \left(r^{o} - r^{n}\right) \qquad (7.11)$$

Equation (7.11) shows that the demographic dividend can also be explained in terms of the growth of the child population, working age population and old age population. When the growth or the relative change of the three population sub-groups is the same, there is no demographic dividend. Notice that $(r^{w} - r^{c})$ is nothing but the relative change or the growth in the child support ratio. Similarly, $(r^{o} - r^{n})$ is the relative change in the proportion of the old population. Finally, $(r^{c} - r^{o})$ is the relative change in the inverse of the ageing index and may be termed as the index of the youngness of the age structure. In other words, the demographic dividend is the result of the increases in the child support ratio, proportion of the old population and the ageing index. It may be noticed that the contribution of the ageing index to the demographic dividend is always negative as the demographic transition is always associated with the decrease in the proportion of the child population and the increase in the proportion of the old population. Thus, the demographic dividend is the sum of the growth of the child support ratio and the proportion of the old population net of the increase in the ageing index.

Table 7.5 presents the average annual growth rate of the total population, child population, working age population and the old age population in India and its states/Union Territories during 2001–2011. In India, the child population in 2011 increased by 2.39% per year whereas the working age population increased by 22.03%. On the other hand, the old age population increased by 30.40% in the year 2011 relative to the year 2001. As a result, the relative increase in the child support ratio was 19.64% in 2011 relative to the year 2001, whereas the increase in the proportion of the old age population was 14.21%. At the same time, the ageing index increased by around 28% during this period so that the ratio of the working age population to the total population increased by 5.84% in 2011 relative to 2001 meaning that the demographic dividend in the country was 5.84% during 2001–2011.

The growth of the child population, adult population and the old population varied widely across states/Union Territories. The child population decreased in 14 states/Union Territories with the decrease being the most rapid in Sikkim. In addition to Sikkim, the child population decreased by more than 10% during 2001–2011 in Andaman and Nicobar Islands and Andhra Pradesh. On the other hand, the child

Table 7.5 Decomposition of the inter-state/Union Territory variance in the change in per capita real GDP

Particulars	Variance-covariance matrix			Contribution to variance in per capita real GDP		Importance to variance in per capita real GDP	
	p	*u*	*d*	Total	%	Total	%
Relative change							
Per capita real GDP (*i*)	0.0345	0.0009	0.0009	0.0362	100.00	0.0468	100.00
Worker productivity (*p*)	0.0359	−0.0021	0.0007	0.0345	95.17	0.0387	82.77
Productive utilisation (*u*)	−0.0021	0.0035	−0.0005	0.0009	2.48	0.0062	13.22
Demographic dividend (*d*)	0.0007	−0.0005	0.0006	0.0009	2.35	0.0019	4.01
Absolute change							
Per capita real GDP (*i*)	327659726	−6201785	9435850	330893790	100.00	380694285	100.00
Worker productivity (*p*)	328767083	−10714775	9607418	327659726	99.02	349089276	91.70
Productive utilisation (*u*)	−10714775	6248339	−1735349	−6201785	−1.87	18698463	4.91
Demographic dividend (*d*)	9607418	−1735349	1563781	9435850	2.85	12906547	3.39

Source Author's calculations

population increased by almost 20% in Manipur during this period. In Meghalaya, Bihar, Jammu and Kashmir, Puducherry, Mizoram, Arunachal Pradesh and Jharkhand also, the child population increased by at least 10% during this period. At the same time, the working age population and the old age population increased in all states/Union Territories. The increase in the working age population was the lowest in Nagaland and the highest in Arunachal Pradesh. The increase in the working age population has also been very low in Kerala and Goa. Finally, increase was the lowest in Nagaland but the highest in Delhi.

As a result of the changes in the age structure of the population, the child support ratio increased in all states/Union Territories but the increase varied widely. In Bihar, Jammu and Kashmir and Nagaland, the child support ratio increased by less than 10%, whereas in Andhra Pradesh, Haryana, Karnataka, Maharashtra, Punjab, Sikkim, Tripura and West Bengal, it increased by at least 25% with the highest increase in Sikkim. Similarly, the proportion of the old population also increased in all states/Union Territories ranging from just around 1% in Arunachal Pradesh to almost 31% in Andaman and Nicobar Islands. The increase in the proportion of the old population was also quite rapid in Andhra Pradesh, Delhi and Goa. Finally, the ageing index also increased in all states/Union Territories; the increase was lowest in Meghalaya but highest in Andaman and Nicobar Island. In Andhra Pradesh, Chandigarh, Delhi, Goa, Haryana, Himachal Pradesh, Karnataka, Maharashtra, Punjab, Sikkim, Uttarakhand and West Bengal also, the ageing index increased by at least 30%.

Table 7.5 also indicates that the relative contribution of the increases in the child support ratio, ageing index and the proportion of the old population to the demographic dividend varies across the states/Union Territories, which means that the variance in the demographic dividend across the states/Union Territories can be decomposed into variance across the states/Union Territories in the increases in the child support ratio, ageing index and the proportion of the old population. Results of this decomposition exercise are presented in Table 7.6. It may be noticed from the table that the covariance between the increase in the child support ratio and the increase in the ageing index is negative. Similarly, the covariance between the ageing index and the proportion of the old population is also negative. As a result, the contribution of the increase in the ageing index and in the proportion of the old population to the variance in the demographic dividend is very small. However, if the sign of the covariance is ignored then Table 7.6 suggests that the variation in the increase in the ageing index across the states/Union Territories is relatively the most important. This is expected as the variation in the increase in the ageing index is determined by both variation in the change in the child population and the variation in the change in the old population. If it is assumed that the increase in the ageing index is divided equally between the increase in the child population and the increase in the old population, the first demographic dividend in the country during 2001–2011 was 5.6%, while the second demographic dividend was only 0.2%. The first demographic dividend was negative in seven states/Union Territories, whereas the second demographic dividend was negative in 16 states/Union Territories. The

Table 7.6 Decomposition of the demographic dividend

Country/State/Union Territory	Relative change during 2001–2011							Demographic dividend		
	Child population	Working age population	Old population	Total population	Child-support ratio	Ageing index	Proportion old	First	Second	Total
India	0.024	0.220	0.304	0.162	0.196	−0.280	0.142	0.056	0.002	0.058
AN Islands	−0.116	0.120	0.381	0.072	0.236	−0.497	0.309	−0.013	0.061	0.048
Andhra Pradesh	−0.113	0.158	0.358	0.097	0.271	−0.471	0.261	0.036	0.026	0.061
Arunachal Pradesh	0.108	0.312	0.243	0.231	0.204	−0.135	0.012	0.136	−0.056	0.081
Assam	0.028	0.223	0.287	0.158	0.195	−0.259	0.129	0.065	−0.001	0.065
Bihar	0.179	0.246	0.337	0.225	0.066	−0.158	0.112	−0.013	0.033	0.021
Chandigarh	0.020	0.196	0.401	0.160	0.176	−0.381	0.241	−0.015	0.051	0.036
Chhattisgarh	0.062	0.278	0.287	0.204	0.216	−0.225	0.082	0.104	−0.030	0.074
Delhi	0.016	0.249	0.467	0.193	0.234	−0.451	0.274	0.008	0.048	0.057
Goa	−0.040	0.083	0.376	0.083	0.123	−0.416	0.293	−0.085	0.085	0.000
Gujarat	0.048	0.219	0.313	0.173	0.171	−0.265	0.140	0.038	0.008	0.046
Haryana	−0.007	0.270	0.326	0.184	0.277	−0.332	0.142	0.111	−0.025	0.087
Himachal Pradesh	−0.060	0.187	0.250	0.123	0.246	−0.310	0.127	0.091	−0.028	0.064
Jammu and Kashmir	0.159	0.238	0.312	0.215	0.079	−0.153	0.097	0.002	0.020	0.022
Jharkhand	0.105	0.242	0.401	0.200	0.137	−0.296	0.201	−0.011	0.053	0.042
Karnataka	−0.050	0.207	0.355	0.145	0.257	−0.405	0.209	0.055	0.007	0.062
Kerala	−0.058	0.056	0.229	0.048	0.114	−0.287	0.181	−0.029	0.038	0.008
Madhya Pradesh	0.044	0.264	0.289	0.187	0.220	−0.245	0.102	0.098	−0.020	0.078
Maharashtra	−0.039	0.215	0.273	0.146	0.254	−0.312	0.127	0.098	−0.029	0.069

(continued)

Table 7.6 (continued)

Country/State/Union Territory	Relative change during 2001–2011							Demographic dividend		
	Child population	Working age population	Old population	Total population	Child-support ratio	Ageing index	Proportion old	First	Second	Total
Manipur	0.198	0.310	0.318	0.275	0.112	−0.120	0.043	0.052	−0.017	0.035
Meghalaya	0.183	0.291	0.273	0.246	0.108	−0.090	0.027	0.063	−0.018	0.045
Mizoram	0.126	0.247	0.336	0.211	0.120	−0.210	0.125	0.015	0.020	0.036
Nagaland	−0.070	0.024	0.129	−0.004	0.095	−0.199	0.133	−0.005	0.033	0.029
Odisha	−0.011	0.182	0.271	0.130	0.193	−0.282	0.141	0.052	0.000	0.052
Puducherry	0.127	0.274	0.396	0.247	0.146	−0.269	0.149	0.012	0.015	0.026
Punjab	−0.073	0.205	0.268	0.132	0.278	−0.341	0.136	0.108	−0.035	0.073
Rajasthan	0.051	0.278	0.294	0.194	0.227	−0.243	0.100	0.106	−0.022	0.084
Sikkim	−0.130	0.227	0.339	0.123	0.357	−0.469	0.216	0.122	−0.018	0.104
Tamil Nadu	0.018	0.179	0.310	0.151	0.162	−0.293	0.159	0.015	0.013	0.028
Tripura	−0.055	0.226	0.219	0.139	0.281	−0.274	0.080	0.144	−0.057	0.087
Uttar Pradesh	0.049	0.260	0.282	0.180	0.212	−0.233	0.101	0.095	−0.015	0.080
Uttarakhand	0.014	0.244	0.320	0.173	0.231	−0.306	0.147	0.078	−0.006	0.072
West Bengal	−0.074	0.207	0.306	0.130	0.281	−0.381	0.176	0.091	−0.014	0.077

Source Author's calculations

Table 7.7 Decomposition of the variation in demographic dividend across the states/Union Territories

Particulars	Variance-covariance matrix			Contribution to variance in demographic dividend		Importance to variance in demographic dividend	
	c	*a*	*o*	Total	%	Total	%
Demographic dividend (*d*)	0.0015	−0.0005	−0.0004	0.0006	100.00	0.0453	100.00
Child support ratio (*c*)	0.0049	−0.0045	0.0011	0.0015	238.25	0.0106	23.33
Ageing index (*a*)	−0.0045	0.0107	−0.0066	−0.0005	−76.08	0.0219	48.25
Proportion old (*o*)	0.0011	−0.0066	0.0051	−0.0004	−62.17	0.0129	28.42

Source Author's calculations

first demographic dividend was relatively the highest in Tripura, whereas the second demographic dividend was the highest in Goa (Table 7.7).

7.9 Discussions and Conclusions

The present analysis confirms that the demographic dividend contributed only marginally to economic growth that India and its constituent states/Union Territories have witnessed during 2001–2011. Moreover, demographic dividend varied widely across the states/Union Territories. Goa is the only state/Union Territory where the demographic dividend actually turned negative during this period and contributed towards slowing down instead of accelerating economic growth. This diversity in demographic dividend across the states/Union Territories of the country is expected as different states/Union Territories of the country are at different stages of demographic transition. It is well known that during the early stage of demographic transition, the child population increases as a result of the decrease in mortality. At the second stage of transition, fertility starts decreasing so that the child population decreases, which results in an increase in the share of the working age population. The first demographic dividend, which is essentially transitory in nature, occurs during this second stage of transition as dependency, either young or old, is low during this phase. The second dividend begins towards the end of the intermediate phase and extends through the late phase only when policies necessary for the second dividend are established (Lee and Mason 2006).

References

Aiyar S, Modi A (2011) The demographic dividend: evidence from the Indian states. International Monetary Fund. Working Paper No. WP/11/38, Washington DC

Bloom DE (2011) Population dynamics in India and implications for economic growth. Program on the Demography of Global Aging. Harvard University

Bloom DE, Williamson JG (1998) Demographic transition and economic miracles in emerging Asia. World Bank Econ Rev 12:419–455

Bloom DE, Finlay JE (2008) Demographic change and economic growth in Asia. Program on the Demography of Global Aging. Harvard University

Bloom DE, Canning D, Sevilla J (2001) Economic growth and demographic transition. Nat Bur Econ Res Working Pap 8685. Cambridge

Bloom D, Canning D, Sevilla J (2003) The demographic dividend: a new perspective on the economic consequences of population change. Santa Monica: RAND. MR-1274

Chandrashekhar CP, Ghosh J, Roychoudhury A (2006) The 'demographic dividend' and young India's economic future. Econ Polit Wkly 41(49):5055–5064

Golley J, Tyers R (2011) Contrasting giants: demographic change and economic performance in China and India. Australian National University. Centre for Applied Macroeconomic Analysis

Government of India (2001) Census of India 2011. Metadata. Registrar General and Census Commissioner, New Delhi

Government of India (2013) Economic Survey 2012–13. Ministry of Finance, New Delhi

Grundy E (1996) Population ageing in Europe. In Coleman D (ed) Europe's Population in 1990s. Oxford University Press, Oxford

Horvitz C, Schemske DW, Caswell H (1997) The relative "importance" of life-history stages to population growth: prospective and retrospective analyses. In: Tuljapurkar S, Caswell H (eds) Structured-population models in marine, terresterial, and freshwater systems. Chapman and Hall, London. Population and Community Biology Series 18: 247–271

Joe W, Dash AK, Agrawal P (2015) Demographic transition, savings, and economic growth in China and India. Institute of Economic Growth. Working Paper No. 351, New Delhi

Kelkar V (2004) India: on the growth turnpike. Narayanan Oration Speech. Australian National University, Canberra

Kim YJ, Strobino DN (1984) Decomposition of the difference between two rates with hierarchical factors. Demography 21(3):361–372

Kumar U (2010) India's demographic transition: boon or bane? A state-level perspective. The conference board. Economics Program Working Paper No. EPWP 10-03, New York

Kumar U, Subramanian A (2012) Growth in India's states in the first decade of the 21st century: four facts. Econ Polit Wkly XLVII(3):48–57

Lee RD, Masson A, Miller T (2000) Life cycle saving and demographic transition in South-East Asia. Population and Development Review 26 (Supplement)

Lee R, Mason A (2006) What is demographic dividend? Finan Develop 30(3)

Mason A (2005) Demographic transition and demographic dividends in developed and developing countries. Paper presented at United Nations expert group meeting on social and economic implications of changing population age structures, Mexico City

Mason A, Lee R, Lee S-H (2008) The demographic transition and economic growth in the Pacific Rim. Paper prepared for the East Asian seminar on economics (EASE), held in Seoul, Korea, 19–21 June 2008

Poorter H, van der Werf A (1998) Is inherent variation in RGR determined by LAR at low irradience and by NAR at high irradience? A review of herbaceous species. In: Lambers H, Poorter H, Van Vuuren MML (eds) Inherent variation in plant growth: physiological mechanisms and ecological consequences, pp 309–336

Preston SH (1994) Population and environment. From Rio to Cairo. International Union for the Scientific Study of Population, Liege

Prskawetz A, Fent T, Barthel W, Crespo-Cuaresma J, Lindh T, Malmbergh B, Halvarsson M (2007) The relationship between demographic change and economic growth in the EU. European Commission, Employment, Social Affairs and Equal Opportunities DG. Research Report No. 32

Ranganathan S, Swain RB, Sumpter DJT (2015) The demographic transition and economic growth: implications for development policy. Palgrave Communication. https://doi.org/10.1057/palcomms.2015.33

Rees M, Grubb PJ, Kelley D (1996) Quantifying the impact of competition and spatial heterogeniety on the structure and dynamics of a four-species guild of winter annuals. Am Nat 147(1):1–32

Rees M, Osborne CP, Woodward FI, Hulme SP, Turnbull LA, Taylor SH (2010) Partitioning the components of relative growth rate: how important is plant size variation? Am Nat 176(6):E152–E161

Singh AK (2016) India's demographic dividend: a sceptical look. Indian J Hum Develop 10(1):10–26

Thakur V (2012) The demographic dividend in India: gift or curse? A state level analysis on differing age structure and its implications for India's economic growth prospects. London School of Economics and Political Science. Development Studies Institute, London. Working Paper No. 12/128

United Nations (2002) World population ageing 1950–2050. United Nations. Department of Economic and Social Affairs, New York

Wang X, Chen K, Huang Z (2013) The impact of China's demographic transition on economic growth and income distribution: CGE modelling with topdown micro-simulation. Paper presented at agricultural & applied economics association's 2013 AAEA & CAES joint annual meeting, Washington, DC, 4–6 August 2013

Wright IJ, Westoby M (2001) Understanding seedling growth relationship through specific leaf area and leaf nitrogen concentration: generalisation across growth forms and growth irradiance. Oecologica 127(1):21-29

Chapter 8
Human Deprivation in Villages

Abstract This chapter analyses the deprivation faced by the village people in India in terms of their capabilities on the basis of a composite human deprivation index which may serve as a yardstick to gauge the extent up to which rural development efforts in the country have contributed towards expanding the capabilities of the village people and, therefore, have contributed to the development of the village. The chapter also explores how the deprivation faced by the village people is influenced by selected defining characteristics of the village. The analysis provides the empirical evidence in support of the village development approach to address the deprivation faced by the village people in the context of their capabilities expansion.

8.1 Introduction

Social and economic transformation of Indian villages has been an abiding concern in India's development discourse right since independence. The reason is obvious. Almost 70% of India's population lives in more than 600 thousand villages of varying population size. By a crude count, more than 75 programmes and schemes have been launched in the country since independence to address the development needs of the village people. The most important of these schemes is the Mahatma Gandhi National Rural Employment Guarantee Act (MGNREGA) which aims at eradicating poverty and promoting sustainability in the rural areas (Government of India 2005). The importance accorded to rural development in India may be judged from the fact that there is a dedicated ministry for rural development in national and state governments and the National Bank for Agriculture and Rural Development (NABARD) has purposely been established to provide the impetus to rural development efforts. The average annual budgetary expenditure on rural economy during 2000–2001 to 2009–2010 was 9.5% of the total combined budgetary expenditure and 2.6% of the gross domestic product of the country at current prices (Jha and Acharya 2011). However, rural development efforts have remained sectoral with different government departments and agencies responsible for different sectors of the rural social and economic landscape. Rural development, in this approach, has a neutral meaning

© Springer Nature Singapore Pte Ltd. 2020

A. R. Chaurasia, *Population and Sustainable Development in India*,
https://doi.org/10.1007/978-981-32-9212-3_8

of development in a strictly geographic context. It is non-linear and highly contextual and has failed to answer territory-specific questions (Prasad 2015).

In recent years, there has been a move towards shifting the focus from rural to village development in India. This shift is reflected through a number of schemes that have been launched recently, including Prime Minister Ideal Village Plan (PMAGY), Member of Parliament Ideal Village Plan (SAGY), Mukhya Mantri Adarsh Gram Yojna of Himachal Pradesh, Smart Village Programme of Andhra Pradesh, Mana Vooru Mana Pranalika Programme of Telangana, etc. These schemes emphasise upon village-specific integrated development. It is argued that being endogenous and self-reliant, development should stem from the heart of the society, which means that development processes should be rooted at the local level, in praxis of each village. This is logical as the richness of development essentially lies in the plurality of its patterns. A focus on village development is all the more pertinent to India as the soul of the country India is argued to reside in villages. A village in India is not merely a place where people live; it has been a design in which the basic values of Indian civilisation are reflected (Beteille 1980). Mahatma Gandhi had argued that India would perish if the village perished and then India would be no more India and her own mission in the world would be lost (Gandhi 1936). He stressed that rebuilding of India could be achieved only by reconstructing villages and visualised a framework of economic self-reliance, social equality and decentralised political system for developing India's villages (Joshi 2002). Rural development efforts in India should therefore be evaluated in the context of the development of Indian villages following the capabilities expansion framework as Sen (1985) argues that development is essentially the expansion of the capabilities of the people.

The present chapter aims at analysing the deprivation faced by the village people in terms of their capabilities on the basis of a composite human deprivation index that has been developed for the purpose. The human deprivation index may also serve as a yardstick to gauge the extent up to which rural development efforts in the country have contributed towards expanding the capabilities of the village people and, therefore, have contributed to the development of the village. The chapter also explores how the deprivation faced by the village people is influenced by selected defining characteristics of the village. The underlying hypothesis is that if the deprivation faced by the village people is contingent upon the defining characteristics of the village, then it is logical to argue that a village-specific development planning and programming should constitute the fundamental building block of the efforts directed towards mitigating the deprivation faced by the village people.

The rest of the chapter is organised as follows: The next section of the chapter describes the data source used in measuring and analysing human deprivation in the villages of the country. We have used the village-level data available through the 2011 population census. Section three of the chapter describes the construction of the human deprivation index used in the present chapter. Section four analyses the variation in human deprivation across villages of the country on the basis of the human deprivation index, whereas section five employs data mining techniques to explore how the human deprivation in the village is influenced by its selected defining characteristics. The last section of the chapter summarises the main findings of the analysis and discusses their policy and programme implications.

8.2 Data Source

The analysis is based on the data available through the 2011 population census in the form of primary census abstract (PCA) and household-level primary census abstract (HLPCA). The PCA is about individuals, while HLPCA is about households in villages in the rural areas and municipal wards in the urban areas of the country. The PCA provides data about total population, population below 7 years of age, social class composition of the population in terms of Scheduled Castes and Scheduled Tribes, number of literates and illiterates in population aged 7 years and above and participation in productive activities. The HLPCA, on the other hand, provides data about household-level living conditions, including material used in the construction of the house, availability of electricity, access to drinking water source and sanitation facilities, availability of kitchen and bathroom, fuel used for cooking, availability of selected household assets, etc. The PCA and the HLPCA are the only sources of data at the village level to assess the deprivation faced by the village people and to link the level of deprivation faced by the village people with the defining characteristics of the village. The PCA has been released after every decennial population census since 1951 but the HLPCA has been released for the first time after the 2011 population census. Combining the village-level data available from the PCA and the HLPCA makes it possible for the first time to analyse human deprivation at the village level in India.

It may be pointed out that a village as defined in the Indian population census is essentially different from the village as a human settlement. A village in the Indian population census has well-defined geographical boundaries and the population of all human settlements within the geographical boundaries is added to obtain the population of the village. There may be a situation where there is no human settlement within the pre-specified geographical boundaries of the village and, in this case, the population of the village is zero. Moreover, no attention is paid to the permanent or the temporary nature of human settlements within the geographic boundaries of the village at the time of population census. A village, as defined in the Indian population census, therefore, is essentially the lowest level administrative unit rather than a human settlement in the conventional sense. There is, however, little information available about the development situation in different human settlements within the geographical boundaries of the same village, and human deprivation situation may vary across human settlements within the same village.

According to the 2011 population census, there were 640,930 villages of varying population size in the country. Out of these, 43,322 villages were uninhabited at the time of the population census, while 41,087 villages were very small, having a population of less than 100 persons, which means that they were essentially hamlets. On the other hand, an exploratory analysis of the village-level data available through

PCA and HLPCA revealed that 26,821 villages in the country had either exceptionally high or low proportion of females to the total population or exceptionally low or high proportion of population aged 7 years and above relative to the total population. Moreover, at least one of the indicators used in the present chapter to construct the human deprivation index could not be calculated for 572 villages because of the inconsistencies in the village-level data available through PCA and HLPCA. All these villages, therefore, have been excluded from the present analysis which covers 529,182 villages of India having a population of at least 100 inhabitants at the time of 2011 population census. These villages accounted for 82.6% of the villages identified at the time of 2011 population census. The total population of these villages was 814,247,606, which is almost 98% of the rural population of the country enumerated at the 2011 population census.

8.3 Human Deprivation Index

We have developed a composite human deprivation index (HDepI) to measure the extent of the deprivation faced by the people. *HDepI* is based on the concept of 'capabilities expansion as development', first propounded by Sen (1985, 1990). Development, in this framework, is the expansion of individual capabilities, broadening the set of valuable beings and doings an individual can achieve. Capabilities, according to Sen, have three domains: endowment, productive capacity and participation opportunity. Deprivation, in this framework, may then be conceptualised in terms of lack of endowment, low productive capacity and absence of opportunities for participation in productive activities. The three domains of deprivation are inter-related and, they, in combination, depict the deprivation faced by the people in the context of capabilities expansion.

Operationalising the capabilities framework to assess the deprivation faced by the people requires selection of indicators to measure endowment, productive capacity and participation opportunity. The selection of indicators may be driven either by the policy perspective or by some theoretical construct or on the basis of the availability of data necessary for estimating the indicator (Hanafin and Brooks 2005). Ideally, all the three approaches should be taken into consideration while selecting indicators. However, the most commonly used approach is the data-driven approach as the availability of data is the most important pre-requisite to estimate indicators reflecting deprivation. We follow, in this chapter, the data-driven approach as the availability of necessary data at the village level is the primary constraint towards measuring human deprivation.

Based on the above considerations and after an examination of the village-level data available through PCA and HLPCA of the 2011 population census, we have used the following set of five indicators to measure the deprivation in terms of endowment, productive capacity and participation opportunity at the village level:

E.1 Proportion of households having none of the seven pre-specified household assets (NAST). Data pertaining to the availability of the following seven household assets were collected at the 2011 population census: (1) radio or transistor; (2) television; (3) computer with or without internet; (4) phone landline or mobile or both; (5) bicycle; (6) moped or scooter or motorcycle or any auto two-wheeler; and (7) care, jeep or wagon. The HLPCA 2011 gives the proportion of households having none of the seven household assets.

E.2 Proportion of households in the village not using banking facilities (NBNK). The proportion of households in the village where at least one member of the household was using the banking facility is available through the HLPCA 2011, which permits estimation of the proportion of households not using banking facilities.

C.1 Population aged 0–6 years as proportion to the total population (P06). The PCA 2011 provides data pertaining to total population and population aged 0–6 years which permit the calculation of the proportion of the population aged 0–6 years. This proportion is very closely related to the expectation of life at birth; the higher the proportion, the lower the expectation of life at birth (Chaurasia 2017). As such, this proportion reflects the deprivation in terms of the health status of the people.

C.2 Proportion of the population aged 7 years and above who cannot read and write with understanding (ILT). This proportion is related to the productive capacity of an individual; the higher the proportion, the lower the capacity to participate in productive activities.

O.1 Average number of days in a year, an individual is not engaged in productive activity (AWW). In the Indian population census, the population is divided into two groups on the basis of their work status—workers and non-workers(Government of India 2011). A person who has worked even for a day during the year prior to the census is classified as a worker. Workers are further classified as main and marginal workers. Main workers are those who had worked for at least 6 months during the year prior to the census. Rest of the workers are classified as marginal workers. Marginal workers are further classified into marginal workers who had worked for 3–6 months and marginal workers who had worked for less than 3 months during the year prior to the census 2011. This means that the entire population of the village can be classified into four mutually exclusive groups on the basis of their work status—population who worked for zero days; population who worked for 0–3 months; population who worked for 3–6 months; and population who worked for 6–12 months during the year prior to the census. These data permit to estimate the average number of days a person worked during the year prior to the census 2011. It is assumed that all workers are engaged at least for 6 months in a year, then the average duration of work in a year may be assumed to be 270 days. The difference between this limiting duration and the actual average number of days of engaged in a year gives a measure of the deprivation in terms of participation opportunity.

The distribution of the five indicators used for the construction of *HDepI* across 529,128 villages of the country is summarised in Table 8.1. The inter-village distribution of the five indicators is essentially different, which implies that there is no single indicator that can capture the multi-dimensional nature of the deprivation faced by the village people. For example, in more than half of the villages of the country, the proportion of households without the specified household assets is very small. On the contrary, in more than half of the villages, the average number of

Table 8.1 Distribution of the indicators used in the construction of the human deprivation index (HDepI) across villages in India

	NAST	NBNK	P06	ILT	AWW
Frequencies					
Very low	52.7	27.1	2.0	17.3	0.1
Low	17.4	10.6	9.5	27.7	0.7
Average	11.4	10.5	19.0	27.0	5.0
High	7.2	10.2	21.4	16.0	13.8
Very high	11.3	41.6	48.1	11.7	80.4
Summary Measures					
Mean	0.242	0.440	0.145	0.332	0.641
SD	0.203	0.296	0.039	0.144	0.124
Skewness	1.287	0.226	0.317	0.725	−0.376
Excess kurtosis	1.472	−1.100	−0.330	0.976	−0.224
Minimum	0.000	0.000	0.031	0.000	0.052
Q1	0.089	0.182	0.116	0.232	0.554
Median	0.188	0.417	0.143	0.316	0.656
Q3	0.339	0.679	0.172	0.414	0.732
Maximum	1.000	1.000	0.258	1.000	1.000

Source Author's calculations based on PCA and HLPCA 2011
Remarks NAST: Proportion of households having none of the seven household assets: (1) radio/transistor; (2) television; (3) computer with or without internet; (4) phone landline or mobile or both; (5) bicycle; (6) auto two-wheeler; and (7) auto four-wheeler. This proportion is very low if NAST < 0.20; low if 0.20 ≤ NAST < 0.30; average if 0.30 ≤ NAST < 0.40; high if 0.40 ≤ NAST < 0.50; and very high if NAST ≥ 0.50. NBNK: Proportion of households not using banking facilities. This proportion is very low if NBNK < 0.20; low if 0.20 ≤ NBNK < 0.30; average if 0.30 ≤ NBNK < 0.40; high if 0.40 ≤ NBNK < 0.50; and very high if NBNK ≥ 0.50. P06: Proportion of population aged 0–6 years to the total population. This proportion is very low if P06 < 0.08; low if 0.08 ≤ P06 < 0.10; average if 0.10 ≤ P06 < 0.12; high if 0.12 ≤ P06 < 0.15; and very high if P06 ≥ 0.15. ILT: Proportion of population aged 7 years and above who cannot read and write with understanding. This proportion is very low if ILT < 0.20; low if 0.20 ≤ ILT < 0.30; average if 0.30 ≤ ILT < 0.40; high if 0.40 ≤ ILT < 0.50; and very high if ILT ≥ 0.50. AWW: Average number of days without work per person as proportion to total working days per person in a year. This number is very low if AWW < 0.24; low if 0.24 ≤ AWW < 0.34; average if 0.34 ≤ AWW < 0.43; high if 0.43 ≤ AWW < 0.53; and very high if AWW ≥ 0.53

days a person is without work is quite high. Similarly, in around one-third of the villages, the proportion of households not using the banking facility is high to very high. Table 8.1 suggests that a multi-dimensional approach should be adopted for measuring and analysing the deprivation faced by the people.

We have used the exploratory factor analysis technique to construct the village-level human deprivation index following the guidelines proposed by OECD (2008). Results of the exploratory factor analysis are presented in Table 8.2. It may be seen from the table that the five indicators of human deprivation used in the present

Table 8.2 Results of the exploratory factor analysis based on the inter-village distribution of five indicators of deprivation

Indicator	Factors		
	I	II	III
Proportion of population aged 0–6 years (P06)	0.842		
Effective illiteracy rate (ILT)	0.852		
Average annual work days (AWW)			0.960
Proportion of households with none of the seven household assets (NAST)		0.646	
Proportion of households not using banking facilities (NBNK)		0.931	
Variance explained	32.05	26.54	22.07
KMO measure	0.584		

Source Author's calculations
Remarks Only those factor loadings are shown which are >0.600

analysis can be grouped into three factors which account for more than 80% of the variation in the original data set. The KMO measure, which is very close to 0.600, and the Bartlett's test of sphericity suggested that the exploratory factor analysis was adequate for combining the five indicators into three factors. The first factor has high loadings in P06 and ILT and accounts for more than 32% of the total variance explained. The second factor has high loadings in NAST and NBNK and accounts for almost 27% of the total variance explained by the factor analysis model. Finally, the third factor has high loadings in AWW and accounts for around 22% of the total variance explained. The exploratory factor analysis thus suggests that the five indicators of human deprivation can be grouped into three factors. Moreover, it is also clear from Table 8.2 that the first factor reflects the productive capacity domain of the capability framework, while the second factor reflects the endowment domain. Finally, the third and the last factor reflects the participation opportunity domain of the capability framework.

Using the results of the exploratory factor analysis and following the approach suggested by Anand and Sen (1997), we have first calculated the endowment deprivation index (*EDepI*), the productive capacity deprivation index (*CDepI*) and participation opportunity deprivation index (*ODepI*) in the following manner:

$$EDepI = (\mathrm{w_A} A^{\alpha} + \mathrm{w_B} B^{\alpha})^{1/\alpha} \tag{8.1}$$

where A is the normalised value of NAST, B is the normalised value of NBNK and $\mathrm{w_A}$ and $\mathrm{w_B}$ are weights such that $\mathrm{w_A} + \mathrm{w_B} = 1$. Similarly,

$$CDepI = (\mathrm{w_P} P^{\alpha} + \mathrm{w_L} L^{\alpha})^{1/\alpha} \tag{8.2}$$

where P is the normalised value of P06, L is the normalised value of ILT and $w_P + w_L = 1$ Finally,

$$ODepI = W \tag{8.3}$$

where W is the normalised value of AWW. Once *EDepI, CDepI* and *ODepI* are calculated, the human deprivation index (*HDepI*) is constructed as the weighted average of order α of *EDepI, CDepI* and *ODepI*:

$$HDepI = (w_E EDepI^{\alpha} + w_C CDepI^{\alpha} + w_O ODepI^{\alpha})^{1/\alpha} \tag{8.4}$$

In the above formulation, α has important impact on *EDepI, CDepI* and *ODepI*, as well as on *HDepI*. When $\alpha = 1$ and $w_A = w_B = 1$, *EDepI* is nothing but the simple average of NAST and NBNK. Similarly, when $\alpha = 1$ and $w_P = w_L = 1$, *CDepI* is nothing but the simple average of P06 and ILT. Finally, when $\alpha = 1$ and $w_E = w_C = T_O = 1$, *HDepI* is nothing but the weighted average of *EDepI, CDepI* and *ODepI*. When $\alpha > 1$, relatively more weight is given to that indicator or that domain in which the deprivation is the highest. This is in conformity with the usual assumption that as the deprivation in any indicator or domain increases, the weight of that indicator or domain on further increase in deprivation in terms of that indicator or domain should also decrease. In other words, when $\alpha > 1$, the *HDepI* attaches more weight to those domains of human deprivation in which the deprivation is relatively higher. The same applies to domain-specific indexes: *EDepI, CDepI* and *ODepI*. The choice of α is, however, arbitrary. Theoretically, α can take any value >1. When $\alpha = 3$, the relative impact of the domain with the highest deprivation is four times the relative impact of the domain with the lowest deprivation. When $\alpha = 5$, the relative impact of the domain with highest deprivation is 16 times higher than the domain with the lowest deprivation. The common practice is to take $\alpha = 3$ to give more weight to that indicator or domain of human deprivation in which the deprivation is the highest. We have also used $\alpha = 3$ in the present chapter to construct *EDepI, CDepI, ODepI* and then to construct *HDepI*.

Calculation of *HDepI* requires estimation of weights w_A and w_B for estimating *EDepI*, weights w_W and w_L for estimating *CDepI* and weights w_E, w_C and w_O for estimating *HDepI* from *EDepI, CDepI* and *ODepI*. We have derived w_A, w_B, w_W and w_L from the factor loadings generated through exploratory factor analysis (OECD 2008). On the other hand, w_E,w_C and w_O have been derived from the proportion of the variance explained by the three factors generated through the exploratory factor analysis. Table 8.3 presents the weights used for the construction of *EDepI, CDepI, ODepI* and *HDepI*.

It is obvious that *HDepI* ranges from 0 to 1; the higher the *HDepI*, the higher the deprivation. Similarly, *EDepI, CDepI* and *ODepI* also range from 0 to 1. Human deprivation may be categorised as very low if *HDepI* < 0.200; low if 0.200 ≤ *HDepI* < 0.300; average if 0.300 ≤ *HDepI* ≤ 0.400; high if 0.400 ≤ *HDepI* < 0.500; and very high if *HDepI* ≥ 0.500. The deprivation in the three domains of human deprivation

Table 8.3 Weights used for the construction of *EDepI*, *CDepI*, *ODepI* and *HDepI*

	EDepI	*CDepI*	*ODepI*
NAST	0.325		
NBNK	0.675		
P06		0.494	
ILT		0.506	
AWW			1.000
HDepI	0.329	0.397	0.274

Source Author's calculations

can also be categorised as very low, low, average, high and very high following the same classification.

8.4 Human Deprivation in Villages

The data available through 2011 population census suggest that human deprivation varies widely across the villages of the country as measured in terms of *HDepI* which is estimated to be the lowest (0.111) in village Palasi of sub-district Hamirpur in district Hamirpur of Himachal Pradesh, but the highest (0.954) in village Lamba of sub-district Kalyansingpur in district Raygada of Odisha. In village Palasi, only around 5% of households was not having any of the seven specified household assets for which data were collected at the 2011 population census. By comparison, there was no household in village Lamba which had at least one of the seven household assets. Similarly, all households in village Palasi were using banking facilities, whereas there was no household in village Lamba which was using the banking facility. More than 88% of the population aged 7 years and above in village Palasi was found to be literate at the time of 2011 population census, whereas the entire population aged 7 years and above in village Lamba was illiterate—unable to read and write with understanding. The population aged 0–6 years constituted only about 12% of the total population of the village Palasi but this proportion was more than 23% in village Lamba. The average number of days a person was without work was only around 36 days (13.3%) out of the maximum of 270 working days in a year in village Palasi, whereas this number was 245 days (90.7%) in village Lamba, indicating that opportunities of participation in productive activities were almost non-existent in the village. The social class composition of the two villages was also different. There was no Scheduled Tribes population in village Palasi, whereas more than 99% of the population in village Lamba was classified as Scheduled Tribes according to the 2011 population census.

The distribution of *HDepI*, *EDepI*, *CDepI* and *ODepI* across 529,128 villages is presented in Table 8.4. There are less than 5% villages where human deprivation may be termed as very low (*HDepI* < 0.200), whereas in almost 50% villages, human deprivation appears to be very high (*HDepI* ≥ 0.500). The kernel density plot of the

Table 8.4 Summary measures of the inter-village distribution of *HDepI*, *EDepI*, *CDepI* and *ODepI*

Particulars	*HDepI*	*EDepI*	*CDepI*	*ODepI*
Frequencies				
Very low (<0.2)	0.50	23.70	2.70	0.10
Low (0.2–0.3)	4.40	13.40	15.00	0.70
Average (0.3–0.4)	15.60	12.90	25.60	5.00
High (0.4–0.5)	29.90	12.20	25.00	13.80
Very high (≥0.5)	49.60	37.80	31.70	80.40
Summary Measures				
Mean	0.535	0.421	0.445	0.621
SD	0.106	0.250	0.142	0.130
Minimum	0.111	0.000	0.013	0.000
Q1	0.463	0.210	0.339	0.530
Median	0.531	0.400	0.431	0.637
Q3	0.604	0.616	0.538	0.717
Maximum	0.954	1.000	0.998	1.000
N	529128	529128	529128	529128
Skewness	0.150	0.253	0.412	−0.376
Excess kurtosis	−0.095	−0.965	−0.198	−0.224

Source: Author's calculations

inter-village distribution of *HDepI* closely resembles a normal distribution with mean 0.535, standard deviation 0.106, skewness 0.150 and excess kurtosis −0.095 but that of *EDepI*, *CDepI* and *ODepI* is essentially different (Fig. 8.1). The distribution of *EDepI* is platykurtic and marginally skewed to the left, that of *CDepI* is highly skewed to the left but closely resembles the normal distribution in terms of kurtosis, whereas the distribution of *ODepI* is highly skewed to the right, although it also closely resembles the normal distribution in terms of kurtosis. It appears that the left-skewed distribution of *CDepI* is compensated by the right-skewed distribution of *ODepI*, whereas the distribution of *EDepI* is more or less normal albeit platykurtic so that the distribution of *HDepI* resembles the normal distribution. Obviously, any discussion based on *HDepI* alone may not be adequate enough to truly reflect the multi-dimensional nature of deprivation faced by people in villages of the country. It is imperative that variation in *EDepI*, *CDepI* and *ODepI* across villages should also be analysed along with the variation across villages in *HDepI*.

In order to analyse simultaneously the variation across villages in *EDepI*, *CDepI* and *ODepI*, we have classified the deprivation in the three domains of human deprivation into three categories—below average, average and above average. If the index is less than 0.300, then the deprivation is categorised as below average. If the index ranges between 0.300 and 0.400, the deprivation is categorised as average and if the index is more than or equal to 0.400, the deprivation is categorised as above average. This categorisation classifies every village into one of the 27 mutually exclusive cat-

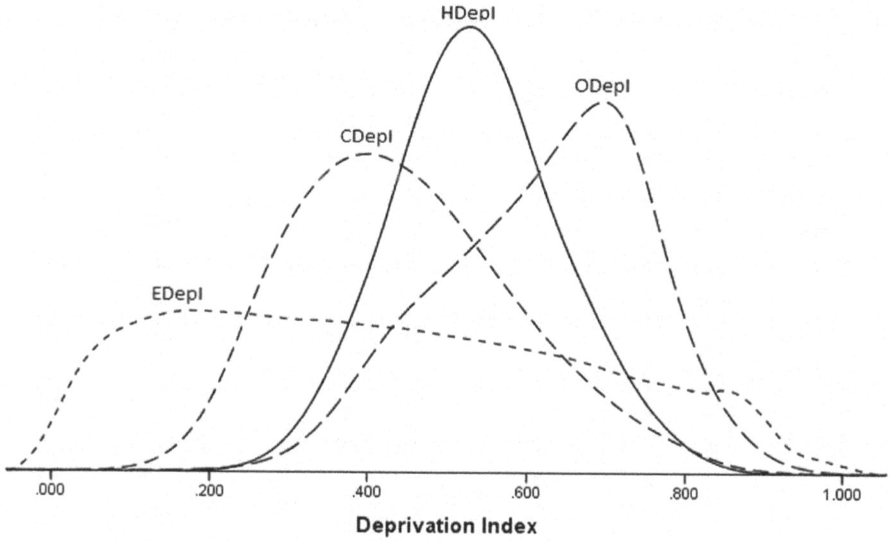

Fig. 8.1 Inter-village distribution of *HDepI*, *EDepI*, *CDepI* and *ODepI* in India

egories on the basis of its *EDepI*, *CDepI* and *ODepI*, as shown in Table 8.5. There are only 740 villages in the country where human deprivation appears to be below the average in all the three domains of human deprivation. In contrast, in almost 30% (158,696) villages of the country, deprivation faced by the people appears to be above the average in all the three domains of human deprivation. On the other hand, the deprivation faced by the people is below the average in the endowment domain, average in the productive capacity domain and above the average in the participation opportunity domain in about 18% (94,727) villages. Similarly, deprivation is above average in the endowment domain, average in the productive capacity domain and above average in the participation opportunity domain in about 11% (57,165) villages of the country. It is obvious from Table 8.5 that there is great diversity across villages in the three domains of human deprivation. This implies that the dynamics of human deprivation is different in different villages.

Any discussion about human deprivation in villages of India is incomplete without a discussion about variation in human deprivation across villages in states/Union Territories of the country. Table 8.6 presents the distribution of villages in different states/Union Territories of the country by the level of *HDepI*. In villages of Bihar, human deprivation appears to be very high as in more than 90% villages, *HDepI* is estimated to be at least 0.500. In contrast, there are less than 4% villages in Kerala, where human deprivation appears to be very high. In addition to Bihar, there are 14 states/Union Territories where human deprivation appears to be very high in at least 50% of the villages. On the other hand, there are only four states/Union Territories including Kerala where human deprivation appears to be very high in less than 10% villages. At the same time, there is no state/Union Territory in the country

Table 8.5 Simultaneous deprivation in the three domains of human deprivation

Sn	Deprivation in			Villages	
	Endowment	Participation capacity	Participation opportunity	Number	Percent
1	Below average	Below average	Below average	740	0.1
2	Below average	Below average	Average	3335	0.6
3	Below average	Below average	Above average	39588	7.5
4	Below average	Average	Below average	426	0.1
5	Below average	Average	Average	3281	0.6
6	Below average	Average	Above average	51087	9.7
7	Below average	Above average	Below average	376	0.1
8	Below average	Above average	Average	2692	0.5
9	Below average	Above average	Above average	94727	17.9
10	Average	Below average	Below average	150	0.0
11	Average	Below average	Average	1067	0.2
12	Average	Below average	Above average	12381	2.3
13	Average	Average	Below average	121	0.0
14	Average	Average	Average	1310	0.2
15	Average	Average	Above average	17612	3.3
16	Average	Above average	Below average	127	0.0
17	Average	Above average	Average	1268	0.2
18	Average	Above average	Above average	34154	6.5
19	Above average	Below average	Below average	426	0.1
20	Above average	Below average	Average	2451	0.5
21	Above average	Below average	Above average	33361	6.3
22	Above average	Average	Below average	456	0.1
23	Above average	Average	Average	4158	0.8
24	Above average	Average	Above average	57165	10.8
25	Above average	Above average	Below average	1340	0.3
26	Above average	Above average	Average	6633	1.3
27	Above average	Above average	Above average	158696	30.0
All				529128	100.0

Source Author's calculations

Table 8.6 Human deprivation in villages of states and Union Territories

Country/State	Human deprivation index (*HDepI*)					Number of villages
	Very low	Low	Average	High	Very high	
Jammu and Kashmir	0.1	0.6	4.3	31.4	63.6	5595
Himachal Pradesh	6.5	19.1	33.9	31.2	9.3	11814
Punjab	0.3	1.8	18.1	63.4	16.4	11513
Chandigarh	0.0	0.0	0.0	100.0	0.0	1
Uttarakhand	1.0	9.2	27.4	38.3	24.1	7921
Haryana	0.1	0.9	16.9	61.8	20.2	6333
Delhi	0.0	0.0	10.1	81.8	8.1	99
Rajasthan	0.1	2.5	19.4	37.9	40.1	39260
Uttar Pradesh	0.1	0.6	5.1	35.8	58.4	89676
Bihar	0.0	0.1	0.8	8.6	90.5	36603
Sikkim	0.5	13.3	43.8	31.0	11.5	384
Arunachal Pradesh	0.0	1.4	10.4	20.7	67.6	2201
Nagaland	0.1	0.8	5.9	15.5	77.8	1196
Manipur	0.2	1.2	3.5	12.5	82.7	1895
Mizoram	0.2	2.7	11.2	16.3	69.7	627
Tripura	0.0	1.1	23.2	47.1	28.6	838
Meghalaya	0.0	0.1	1.8	9.3	88.8	4346
Assam	0.1	1.6	9.3	26.5	62.5	22433
West Bengal	0.1	0.9	11.6	30.6	56.8	35189
Jharkhand	0.0	0.3	3.3	18.1	78.2	25627
Odisha	0.1	0.9	8.5	30.0	60.5	39345
Chhattisgarh	0.1	4.7	16.7	26.9	51.5	18082
Madhya Pradesh	0.0	1.0	8.5	25.6	64.8	47427
Gujarat	0.3	4.8	25.0	34.6	35.3	17168
Daman and Diu	0.0	5.9	64.7	29.4	0.0	17
Dadra and Nagar Haveli	0.0	0.0	1.7	25.4	72.9	59
Maharashtra	1.4	20.4	40.3	24.1	13.8	38029
Andhra Pradesh	0.2	6.8	35.6	32.8	24.6	23988
Karnataka	2.9	16.5	34.6	28.1	17.9	25223
Goa	0.0	3.1	54.9	36.9	5.1	293
Lakshadweep	0.0	0.0	0.0	66.7	33.3	3
Kerala	0.0	2.8	51.7	41.7	3.7	992
Tamil Nadu	0.8	10.1	31.5	36.1	21.6	14622

(continued)

Table 8.6 (continued)

Country/State	Human deprivation index (*HDepI*)					Number of villages
	Very low	Low	Average	High	Very high	
Puducherry	0.0	1.1	31.1	58.9	8.9	90
Andaman and Nicobar Islands	0.0	3.8	40.6	42.3	13.4	239
India	0.5	4.4	15.6	29.9	49.6	529128

Source Author's calculations

where human deprivation appears to be very low in more than 10% villages in the state/Union Territory on the basis of *HDepI*. In fact, there is not a single village in 15 states/Union Territories of the country where human deprivation appears to be very low in the sense that *HDepI* < 0.200.

Regional pattern in human deprivation across villages is also evident from Table 8.6. Throughout the central and the eastern part of the country, human deprivation appears to be very high in majority of the villages. In contrast, the proportion of villages where human deprivation is very high is relatively low in the southern part of the country. In the northern part of the country, human deprivation appears to be neither very high nor very low in most of the villages. The regional pattern in human deprivation at the village level is reflected in the regional pattern in human deprivation across the states/Union Territories of the country (Fig. 8.2).

The relative contribution of the variation in *EDepI*, *CDepI* and *ODepI* across villages to the variation in *HDepI* also varies across the states/Union Territories. In most of the states/Union Territories, the main contributor to the variation in *HDepI* is the variation in deprivation in endowment domain as measured through *EDepI*. However, in Himachal Pradesh, Uttarakhand, Goa and Kerala, the main contributor to the variation in *HDepI* is the variation in *ODepI*. On the other hand, variation across villages in *CDepI* is not a dominant contributor to the variation across villages in *HDepI* in any state/Union Territory of the country.

8.5 Classification of Villages

Are the defining characteristics of the village related to the level of human deprivation in the village? We have used the classification modelling approach (Tan et al. 2006; Han et al. 2012) to analyse the relationship between selected defining characteristics of the village with the extent of the deprivation faced by the people of the village. Classification modelling involves classifying villages on the basis of their defining characteristics and then analysing the distribution of *HDepI* in different groups of villages. The classification and regression tree (CRT) method (Breiman et al. 1984) has been used for the purpose. CRT is a non-parametric method that divides villages into different groups in such a way that within-group homogeneity with respect to

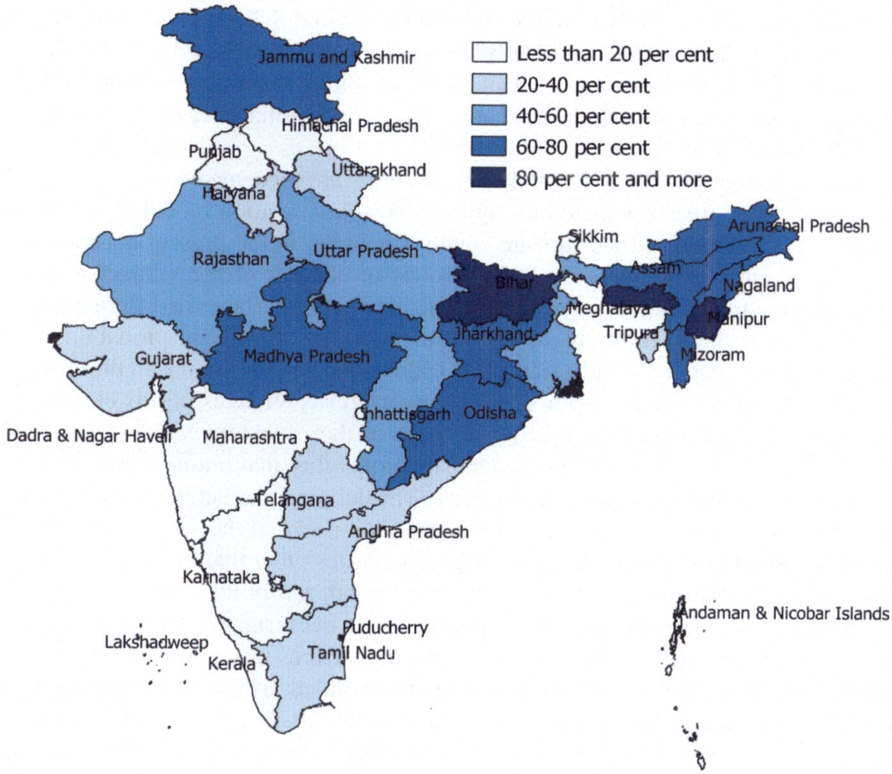

Fig. 8.2 Inter-state/Union Territory distribution of the proportion of villages with very high *HDepI*

HDepI is the maximum. The technique sorts villages into mutually exclusive groups based on that independent variable which causes the most effective split. The process is repeated until either the perfect homogeneity is achieved or a pre-decided stopping criterion is met (Ambalavanan et al. 2006; Lemon et al. 2003). A group in which all villages have the same value of the classification variable is termed as 'pure'. If a group is not 'pure', then the impurity within the group can be measured by a number of coefficients. We have used the Gini coefficient of impurity to measure the degree of homogeneity within groups identified through classification modelling. If the dependent variable is categorical or ordinal, the method gives the distribution of the dependent variable across villages in each group or category. If the dependent variable is continuous, the method gives estimates of arithmetic mean and standard deviation of the dependent variable in each group or category of villages.

For the purpose of classification modelling, we have taken into consideration three defining characteristics of the village—size of the village, its social class composition and the gender balance of the village population. The size of the village is measured in terms of the population enumerated at the 2011 population census. On the other hand, the social class composition of the village is measured by two indicators:

(1) proportion of castes other than Scheduled Castes and Scheduled Tribes; and (2) proportion of Scheduled Tribes. Finally, the gender balance of the village population is measured in terms of the proportion of females to the total population of the village. The selection of these three defining characteristics of the village was primarily based on the availability of data at the village level.

Results of the classification modelling exercise are presented in Table 8.7 and the classification tree is depicted in Fig. 8.3. The classification modelling exercise suggests that 529,128 villages of the country included in the present analysis can be grouped into 10 mutually exclusive groups or clusters of the villages and the distribution of the human deprivation index *HDepI* across villages in different group or cluster is essentially different. In villages included in cluster 1, the probability that human deprivation is very high ($HDepI \geq 0.500$) is the highest among all the clusters identified through the classification modelling exercise. Villages of this cluster are almost entirely dominated by Scheduled Tribes as they constitute at least 97.6% of the population of villages in this cluster. The probability that human deprivation is very high is also high in villages of cluster 2. The defining characteristics of villages included in this cluster are relatively large population size of the village (>900) and near total population of the village belonging to castes other than Scheduled Castes and Scheduled Tribes (>99.8%). On the other hand, the probability of very high human deprivation is relatively the lowest in villages of cluster 10. The proportion of castes other than Scheduled Castes and Scheduled Tribes in villages of this cluster ranges between 42.7 and 99.8%; proportion of Scheduled Tribes is less than or equal to 29.7% and the proportion of females in the population is more than 48.5%, which suggests that the gender balance in villages of this cluster is generally favourable to females. It is obvious from Table 8.7 and Fig. 8.3 that the defining characteristics of the village has substantial influence on the deprivation faced by the people of the village. These defining characteristics of the village, however, are, essentially, exogenous to the social and economic development system. This means that social and economic development processes are not efficient and effective enough to address the deprivation faced by the people living in villages.

8.6 Conclusions

It is clear from the present analysis that human deprivation is quite rampant in India's villages and there are only a handful of villages where human deprivation can be termed as low. These observations question the effectiveness of rural development efforts in addressing the development and welfare needs of the village people in the context of the expansion of their basic capabilities—endowment, productive capacity and participation opportunity.

There appears a distinct regional pattern in the proportion of villages with very high human deprivation. In majority of the villages in the central and eastern part of the country, human deprivation is high to very high, whereas in the southern part of the country, the proportion of villages with very high human deprivation is relatively

Table 8.7 Results of the classification modelling exercise

Cluster	Village characteristics				Probability that HDI in the village of the cluster is					Number of villages in the cluster
	Proportion other castes	Scheduled tribes	Proportion females	Population size	Very low	Low	Average	High	Very high	
1	≤0.161	>0.976			0.001	0.012	0.057	0.134	0.797	29681
2	>0.998			>900	0.002	0.014	0.066	0.195	0.724	10292
3	≤0.161	≤0.976			0.003	0.023	0.091	0.221	0.662	41699
4	>0.161, ≤0.427	>0.587			0.002	0.028	0.123	0.261	0.587	19127
5	>0.998			≤900	0.012	0.051	0.138	0.265	0.534	32300
6	>0.161, ≤0.427	≤0.587			0.003	0.031	0.141	0.311	0.514	33907
7	>0.427	≤0.002	≤0.485		0.004	0.028	0.113	0.345	0.511	115337
8	>0.427, ≤0.998	>0.297	>0.485		0.003	0.038	0.171	0.311	0.477	13866
9	>0.427	>0.002	≤0.485		0.004	0.060	0.197	0.311	0.426	61842
10	>0.427, ≤0.998	≤0.297	>0.485		0.007	0.065	0.219	0.324	0.385	171077
All					0.005	0.044	0.156	0.299	0.496	529128

Source Author's calculations

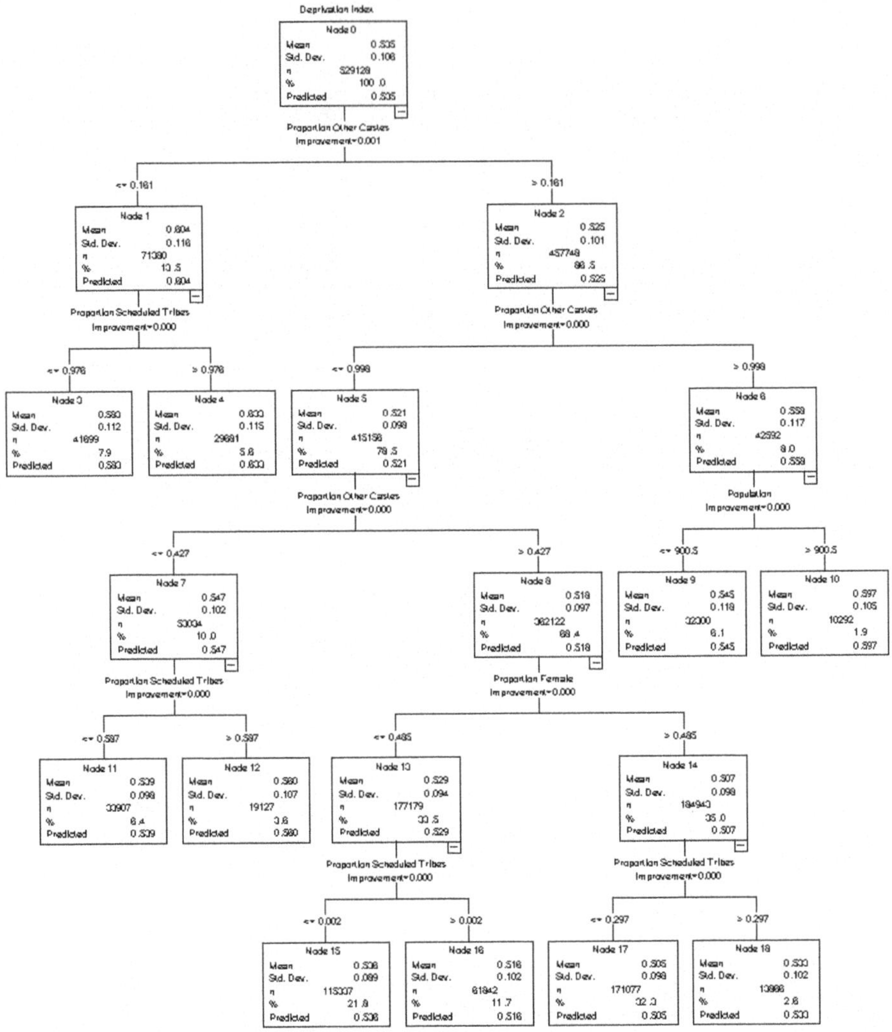

Fig. 8.3 Classification of villages on the basis of *HDepI* and selected defining characteristics of the village

low. In the northern part of the country, on the other hand, human deprivation is neither very high nor very low in majority of the villages.

Among different states/Union Territories of the country, the challenge of very high human deprivation in villages appears to be alarming in Bihar, Jharkhand and Madhya Pradesh and in almost all the north-eastern states of the country. In more than 60% villages of these states, human deprivation appears to be very high.

The size of the village population, its social class composition and the gender balance of the village population have been found to have a strong influence on the deprivation faced by the village people. These characteristics of the village are essentially exogenous to the social and economic development system. This means that the conventional approach to rural development has not been able to make a telling impact upon the deprivation faced by the village people.

The primary reason behind high to very high deprivation being faced by the village people appears to be very limited opportunities for the village people for participation in productive activities. In more than four-fifths villages of the country, deprivation in the opportunity domain of the capabilities framework is found to be very high. Improving the opportunities for productive participation of the village people may go a long way towards expanding capabilities of village people, thereby addressing human deprivation in the villages of the country.

The volatility in deprivation in the endowment domain of the capability framework is very high across the villages of the country. This indicates towards substantial inequality in the distribution of material resources across the villages of the country.

An important observation of the present analysis is that the three domains of the capabilities framework—endowment, productive capacity and participation opportunity—are not essentially inter-linked at the village level. There may be many reasons for this observed lack of inter-dependence in the three domains of capabilities. For example, village people may not have sufficient endowment for enhancing their productive capacity or village people may not have enough participation opportunities for utilising their productive capacity to increase their endowment. There may also be a situation where the village people have the participation opportunity but their productive capacity is too poor to take the advantage of the opportunities available to them for raising their endowment. There is a need to carry out village-specific analysis of the observed disconnect among the three domains of capabilities expansion.

The present analysis provides the empirical evidence in support of the village development approach to address the deprivation faced by the village people in the context of their capabilities expansion. It is imperative that any approach towards addressing human deprivation at the village level must take into account village-level endowments, should enhance the village productive capacity and must increase opportunities for productive participation of the village people. The present analysis, however, suggests that the necessary village-level institutions for enhancing the productive capacity of the village people and for increasing the opportunities of their participation in productive activities are largely absent in the country.

References

Ambalavanan N, Carlo WA, Shankaran S, Ban CM, Emrich SL, Higgins RD, Tyson JE, O'Shea TM, Laptook AR, Ehrenkranz RA, Donovan EF, Waleh MC, Goldberg RN, Das A, National Institute of Child Health and Human Development Neonatal Research Network (2006) Predicting outcomes of neonates diagnosed with hypoxemic-ischemic encephalopathy. Pediatrics 118:2084–2093

Anand S, Sen AK (1997) Concept of human development and poverty: a multidimensional perspective. New York, United Nations Development Programme. Human Development Chapters

Beteille A (1980) The Indian villages: past and present. In Hobsbaun EJ, Kula W, Mitra A, Raj KN, Sachs I (eds) Peasants in history: essays in honour of Daniel Thorner. Oxford University Press, Calcutta

Brieman L, Friedman JH, Olshen RA, Stone CJ (1984) Classification and regression trees. CRC Press, New York

Chaurasia AR (2017) Human development in Gujarat. Evidence from 2011 population census. Indian J Hum Dev 11:1–21

Gandhi MK (1936) Harijan 63:241

Government of India (2005) The National Rural Employment Guarantee Act, 2005. The Gazette of India 48, Part II Section 1

Government of India (2011) Population census 2011. Metadata. Registrar General and Census Commissioner, New Delhi

Han J, Kamber M, Pei J (2012) Data mining. Concepts and techniques. Elsevier, Amsterdam

Hanafin S, Brooks A-M (2005) Report of the development of a national set of child well-being indicators in Ireland. The National Children's Office, Dubkin

Jha P, Acharya N (2011) Rural economy in India's budgets since 1950s: an assessment. Rev Agrar Stud 1:134–156

Joshi D (2002) Village economy. Gandhiji on villages. Gandhi Book Centre, Mumbai

Lemon SC, Roy J, Melissa A, Clark PD, Friedman WR (2003) Classification and Regression Tree analysis in public health: methodological review and comparison with logistic regression. Ann Behav Med 26:172–181

OECD (2008) Handbook on constructing composite indicators. Methodology and user guide. Organisation of Economic Cooperation and Development, Paris

Prasad RR (2015) Rural development versus village development. http://www.governancenow.com/views/columns/rural-developmentversus-village-development. Last accessed on 10 Apr 2017

Sen AK (1985) Commodities and capabilities. Oxford University Press, Oxford

Sen AK (1990) Development as capability expansion. In: Griffin K, Knight J (eds) Human development and international development strategy for the 1990s. Macmillan, London

Tan P-N, Steinbach M, Kumar V (2006) Introduction to data mining. New Delhi, Pearson Education

Chapter 9
Child Deprivation

Abstract This chapter analyses regional, residence and social class variation in child deprivation in India. The underlying premise of the chapter is that deprivation should be measured in terms of services and facilities that need to be provided to children to ensure their survival, growth, development and protection. Child deprivation in India remains highly pervasive. A substantial proportion of Indian children are devoid of even the basic services and facilities necessary for their survival, growth, development and protection. Improving the needs effectiveness and increasing the administrative capacity of child well-being efforts may contribute substantially towards addressing the deprivation faced by the children of the country.

9.1 Introduction

Concerns for child well-being in India are enshrined in the fundamental rights and the directive principles of state policy as inscribed in the Constitution of India. These concerns are amply reflected in the National Policy on Children, announced, for the first time, in 1974 (Government of India 1974). The policy commits provision of adequate services to children, both before and after birth and through the period of growth, to ensure their full physical, mental and social development. It also emphasises that the state will progressively increase the scope of such services so that, within a reasonable time, all children in the country enjoy optimum conditions for their survival, balanced growth and cognitive development. In 1992, India ratified the United Nations Convention on the Rights of the Child (United Nations 1989) and committed herself to protecting and advancing children's rights. Since then, there have been many attempts to main streaming child rights issues in the development agenda of the country through policy and programme initiatives. The latest in these attempts is the National Policy on Children 2013, which affirms the commitment of the Government of India to rights-based approach to addressing the continuing and emerging challenges in promoting and sustaining child well-being (Government of India 2013). The policy recognises that survival, health, nutrition, development, education, protection and participation are the undeniable rights of every child of the country. The Government of India has also launched the Integrated Child Protec-

© Springer Nature Singapore Pte Ltd. 2020

A. R. Chaurasia, *Population and Sustainable Development in India*,
https://doi.org/10.1007/978-981-32-9212-3_9

tion Scheme (ICPS) which is based on the cardinal principles of 'protection of child rights' and 'best interests of the child'. It aims to promote the best interests of the child and prevent violations of child rights through appropriate punitive measures against perpetrators of abuse and crimes against children and to ensure rehabilitation for all children in need of care and protection by creating a protective environment through improving regulatory frameworks, strengthening structures and professional capacities at national, state and district levels so as to cover all child protection issues and provide children-friendly services at all levels (Government of India 2007).

Despite repeated commitments and policy prescriptions, protecting child rights in India in terms of their survival, growth, cognitive development and protection from social, economic, cultural and environmental vulnerabilities remains a major development challenge. Although child protection is now increasingly being recognised as an important, indispensable, dimension of human development, yet, children are particularly vulnerable to the deprivation of their specific needs simply because they are not full social and economic agents (White et al. 2002). They are exposed more to household, societal, economic and cultural vulnerabilities than adults and their dependence on public provision of goods and services is definitely heavier than other members in the family and the society (Gordon et al. 2003a, b; Minujin et al. 2005; Notten and de Neubourg 2011; Waddington 2004; White et al. 2002). It is well known that deprivation of children from services and facilities that they need for their survival, growth, cognitive development and protection contributes to child poverty and this deprivation is sensitive to prevailing social, economic, cultural and environmental milieu.

Child deprivation, by its very conceptualisation, is multi-dimensional simply because child well-being is essentially a multi-dimensional concept. There are many dimensions of child well-being and it is common sense to assume that a child may experience deprivation in more than one dimension of child well-being. However, there are only a few studies in India that have attempted to analyse and highlight the challenge of child deprivation through a multi-dimensional perspective (Citizen's Initiative for the Rights of Children under Six 2006; Dréze et al. 2007; Chandrashekhar and Suryanarayana 2007; Chaurasia 2010).

This chapter analyses regional, residence and social class variation in child deprivation in India on the basis of a composite child deprivation index. The underlying premise of the chapter is that deprivation should be measured in terms of services and facilities that need to be provided to children to ensure their survival, growth, development and protection. It is argued that a child is deprived if it fails to receive a service or if it does not have access to a facility for which it has the right. This conceptualisation emphasises that it is the needs effectiveness and capacity efficiency of the service provider that is primarily responsible for different forms of deprivation that children face.

The chapter follows the rights framework which has been evolved in recognition of the fact that a child is a human being and, therefore, has certain moral status. This implies that there are certain things that should not be done and there are certain things that should be done to children. The rights that serve the 'best interests' of children can broadly be grouped into positive rights and moral rights. Positive rights are recognised by law. Moral rights are based on some moral theory but are largely accepted by the society. The important point is that entailing positive rights does not always ensure moral rights, although it is argued that possession of positive rights is sufficient to outweigh all moral considerations (Nozick 1974). The counter argument is that mere possession of positive rights may not out-balance every other moral claim of children because they lack the capability to exercise positive rights. There must be conditions in place, to ensure that positive rights are translated into moral rights.

The chapter is organised as follows. The next section describes the data source. The chapter is based on the data available through the Rapid Survey of Children conducted by the Government of India 2013–2014. The third section describes the child deprivation index that constitutes the basis for deprivation analysis. The fourth section presents estimates of the child deprivation index for India as a whole for the total population and for different population subgroups while section five presents the estimates of the child deprivation index for the states of the country by residence and social class. Section six analyses residence and social class inequalities in child deprivation. The last section of the chapter summarises the findings of the analysis and discusses the policy and programme implications.

9.2 Data Source

The chapter is based on the data available through the Rapid Survey of Children (RSoC) 2013–2014 which was conducted by the Government of India, Ministry of Women and Child Development towards strengthening the data system on children and women (Government of India 2015). RSoC 2013–2014 covered 105,843 households in 29 states of the country. Union Territories were not covered in the survey. In the first stage of sample selection, primary sampling units (PSU) were selected through probability proportional to size with replacement, and within each PSU, households were selected using the circular systematic sampling procedure. Respondents of the survey included head of the household surveyed, all ever married women aged 15–49 years in the household and all currently married women aged 15–49 years in the household surveyed who were pregnant at the time of the survey. Based on the data collected during the survey, key indicators reflecting the situation of children were estimated and released as fact sheets. These fact sheets allow measurement of child deprivation in the country and in its constituent states and permit analysis of residence and social class inequalities in child wellbeing.

9.3 Child Deprivation Index

The conceptual model of child deprivation is based on the idea of distinct domains of deprivation. Children may be counted as deprived in one or more of these domains depending on the number and types of deprivation they experience. This approach avoids the need to make judgments about the complex link between different types of deprivation. The domain-based framework of measuring and analysing child deprivation essentially stems from two considerations. First, the different domains of child well-being have different importance in different ages of the childhood period. For example, during the first 5 years of life, the most important considerations for child well-being are related to survival and growth of the child. By comparison, during 6–14 years of age, the key consideration related to child well-being is the cognitive development of the child. Finally, protection may attract top priority in considerations related to child well-being during 15–18 years of age.

The second rationale for adopting a domain-based approach of analysing child deprivation rests in the fact that interventions directed to addressing domain-specific issues and concerns are essentially different. Thus, interventions related to enhancing survival have little or limited bearing on issues and concerns related to growth. Similarly, interventions related to ensuring normal growth have little to do with interventions related to cognitive development of the child. Finally, protecting the child from environmental and societal hazards may require entirely different types of interventions.

The foregoing considerations suggest that at least four domains of child well-being can readily be identified for analysing child deprivation: (1) the survival domain; (2) the growth domain; (3) the development domain; and (4) the protection domain. In addition, environmental conditions may also influence child well-being and therefore the environment domain may also be taken into consideration in any analysis of child deprivation. The analytical framework used in the present analysis focuses on these five domains of child well-being. The analytical strategy is to first measure the deprivation in each of the five identified domains of child well-being through suitably design deprivation index, and then combine domain-specific deprivation indexes into one, single, index of child deprivation. The approach recognises that child deprivation is essentially multi-dimensional and a child may experience deprivation in one or more domains of child well-being.

Child well-being in different domains can be measured in terms of outcomes and outputs of domain-specific interventions. This approach of measuring child well-being is directed towards ensuring that every child receives services and has access to facilities that the child needs for its survival, growth, development and protection. Deprivation, in this conceptualisation, is measured in terms of coverage of services and access to facilities; and poor outcomes in different domains may be a reflection of incomplete coverage of services and/or lack of availability of and therefore access to facilities. An advantage of this approach is that it helps in identifying priorities of child well-being that can constitute the evidence base for planning and programming efforts and interventions directed towards mitigating child deprivation.

The next step in the construction of the child deprivation index is the selection of interventions in each domain. It is logical to assume that there may be more than one intervention in each domain that have an impact on child well-being. There is, however, no pre-set rule for the selection of interventions. In general, selection of interventions is primarily subjective, depending upon the availability of the data necessary to measure the coverage of or access to the intervention to reflect deprivation.

An important consideration in this regard is that data should be available through population-based, statistically representative, survey so as to estimate coverage and access.

The Government of India has released country and state fact sheets on children on the basis of RSoC 2013–2014 (Government of India 2015). These fact sheets give estimates of indicators related to well-being of children below 6 years of age and their mothers covering aspects of child development, maternal care, school/college attendance among persons aged 5–24 years, early childhood care and the enabling environment, like access to drinking water and use of toilet facilities. Out of the indicators given in the fact sheets, the following 10 interventions were selected for analysing deprivation in five domains of child well-being (Table 9.1). The list of indicators cannot be termed as comprehensive or complete because estimates of some

Table 9.1 Indicators selected for measuring child deprivation

Domain of child well-being	Indicator	Description
Survival	1.1 Full antenatal care	Proportion of women who did not receive full antenatal care during their last pregnancy (S_1)
	1.2 Check-up of the new born	Proportion of newborn who did not receive first check up within 24 h of birth/discharge from the hospital (S_2)
	1.3 Birth weight	Proportion of newborn who weighed less than 2.5 kg at birth (S_3)
	1.4 Immunisation	Proportion of children 12–23 months of age not fully immunised (S_4)
Growth	2.1 Initiation of breast feeding	Proportion of children 0–23 months of age who were not initiated breast feeding within one hour of birth (G_1)
	2.2 Linear growth	Proportion of children aged 0–59 months who were stunted (G_2)
	2.3 Parenteral growth	Proportion of children aged 0–59 months who were wasted (G_3)
Development	3.1 Pre-school education	Proportion of children aged 3–6 years who were not attending preschool education (D_1)
Protection	4.1 Early marriage	Proportion of girls aged 10–19 years ever married (P_1)
Environment	5.1 Sanitation	Proportion of households practising open defecation (E_1)

of the commonly used indicators of deprivation are missing simply because estimates are not currently available through RSoC 2013–2014. For example, estimates of the proportion of children aged 6–14 years in school are not currently available through RSoC 2013–2014. Similarly, prevalence of child labour is not available. Finally, an important component of the environment domain is the access to safe drinking water sources, but the definition of the safe drinking water is not clear in RSoC 2013–2014.

An important consideration in the selection of the indicators for measuring and analysing child deprivation is that indicators used in the construction of the child deprivation index must not be multi-collinear. If two indicators are multi-collinear then there is no justification of including both the indicators in the construction of the child deprivation index as the change in one indicator automatically induces change in the second indicator. Testing of multi-collinearity among the indicators used in the construction of the child deprivation index is therefore important.

Table 9.2 presents simple zero-order correlation coefficients among the indicators used in the construction of the child deprivation index. These correlation coefficients are based on inter-state variation as revealed through RSoC 2013–2014. The table shows that only 17 of the 45 correlation coefficients are statistically significant and there is no correlation coefficient which is more than 0.75 or less than −0.55. This suggests absence of multi-collinearity among the indicators used in the construction of the child deprivation index.

The next step in constructing the child deprivation index is to normalise all indicators so that they vary between 0 and 1. Normalisation can be absolute or relative. Absolute normalisation is based on fixed goal posts which reflect the minimum and maximum possible values of the indicator. Relative normalisation, on the other hand, is based on the data set. One disadvantage of relative normalisation is that the administrative unit having relatively the lowest value of an indicator will be characterised as the administrative unit with no deprivation while the administrative unit with the highest value of the indicator will be characterised as the administrative unit with total deprivation which may be actually not the case. In the deprivation analysis, therefore, it is more appropriate to normalise indicators in the absolute sense rather than in the relative sense. Since all indicators used in the present analysis are proportions, the possible minimum and maximum value of all indicators is 0 and 100, respectively.

The last but one step in the construction of the child deprivation index is to calculate the deprivation index for each domain separately. For this purpose, the mean of order 3 was calculated. Thus, the deprivation index in the survival domain (SDI) was calculated as:

$$\mathrm{SDI} = \sqrt[3]{\left(S_1^3 + S_2^3 + S_3^3 + S_4^3\right)/4} \tag{9.1}$$

Similarly, the deprivation index in the growth domain (GDI) is calculated as:

$$\mathrm{GDI} = \sqrt[3]{\left(G_1^3 + G_2^3 + G_3^3\right)/3} \tag{9.2}$$

Table 9.2 Simple zero-order correlation coefficients among indicators used in the construction of the deprivation index

Indicator	S_1	S_2	S_3	S_4	G_1	G_2	G_3	D_1	P_1
S_2	*0.432*								
S_3	*0.474*	−0.227							
S_4	*0.629*	*0.435*	0.000						
G_1	*0.521*	0.090	*0.568*	−0.007					
G_2	*0.664*	0.150	0.255	*0.466*	*0.447*				
G_3	−0.134	*−0.544*	0.246	−0.102	−0.235	−0.030			
D_1	*0.546*	*0.409*	0.088	*0.700*	−0.002	0.272	−0.130		
P_1	0.344	0.185	0.192	*0.450*	−0.064	0.010	0.076	*0.575*	
E_1	*0.487*	−0.093	0.335	0.261	0.237	*0.746*	0.303	0.113	0

Source Author's calculations

On the other hand, there is only one indicator in development, protection and environment domains so that deprivation indexes in development domain (DDI), protection domain (PDI) and environment domain (EDI) were calculated as

$$\mathrm{DDI} = D_1 \quad (9.3)$$

$$\mathrm{PDI} = P_1 \quad (9.4)$$

$$\mathrm{EDI} = E_1 \quad (9.5)$$

Finally, the five domain-specific deprivation indexes were combined to obtain the child deprivation index:

$$\mathrm{CDI} = \sqrt[3]{\left(\mathrm{SDI}^3 + \mathrm{GDI}^3 + \mathrm{SDI}^3 + \mathrm{SDI}^3\right)/4} \quad (9.6)$$

It is obvious that CDI varies between 0 and 1. When CDI = 0, there is no deprivation in any of the five domains and in any of the 10 indicators of child well-being used in the present analysis. On the other hand, when CDI = 1, there is total deprivation in all five domains and in all the 10 indicators of child well-being. This means that the higher the CDI, the higher the deprivation. Similarly, the domain deprivation indexes also vary between 0 and 1. If the deprivation index in a domain of child well-being is 0, then this implies that there is no deprivation in the domain. On the other hand, if the deprivation index is 1 in a domain, then this means that there is total deprivation in that domain of child well-being. A comparison of domain-specific deprivation indexes with the child deprivation index helps in identifying the domain, which is largely responsible for the given level of child deprivation.

Based on the level of the child deprivation index (CDI), child deprivation may be categorised as very low if the CDI is less than 0.3, low if the CDI ranges between 0.3 and 0.4, medium if the CDI ranges between 0.4 and 0.5, high if the CDI ranges between 0.5 and 0.6, and very high if the CDI ranges between 0.6 and 0.7. Finally, child deprivation may be termed as extreme if the CDI is 0.7 and above. The same classification may be applied to characterising deprivation in different domains of child well-being.

It may, however, be emphasised that the CDI and domain-specific deprivation indexes presented in this chapter are sensitive to the conceptualisation of child well-being and the indicators selected to measure the child well-being. The CDI also depends upon how different indicators are grouped into different domains of child well-being. When indicators of child well-being are changed or when their grouping in different domains of child well-being is modified, the domain-specific deprivation indexes and hence the CDI are also changed. This limitation is, however, omnipresent in every empirical analysis.

9.4 Child Deprivation in India

Estimates of the child deprivation index (CDI) for India as a whole and separately for rural and urban population as well as for Scheduled Castes (SC), Scheduled Tribes (ST), Other Backward Classes (OB) and Other classes (OT) are presented in Table 9.3, along with deprivation indexes in different domains of child well-being. For India as a whole, the CDI is estimated to be 0.433, which suggests that child deprivation in India is quite pervesive and has persisted over time. The evidence available from RSoC 2013–2014 indicates that the country has to go a long way to achieve the cherished goal of ensuring that every child of the country has access to necessary facilities and receive basic services to which it has the right. Table 9.3 also shows that there is a very substantial gap in child deprivation across rural and urban areas of the country. Similarly, child deprivation varied by social class. The strength of the rural–urban and social class differentials in CDI suggests that residence and social class disparities in child deprivation have also persisted over time.

Among different domains of child well-being, deprivation appears to be the highest in the survival domain followed by environment and growth domains of child well-being but the lowest in the protection domain. Residence and social class differentials in deprivation in different domains of child well-being also vary widely. In the survival domain of child well-being, deprivation in the rural areas is substantially higher than the deprivation in the urban areas. Among different social classes, child deprivation in the survival domain is the highest in Scheduled Tribes but the lowest in Other social classes. Child deprivation is also quite substantial in the growth domain but residential and social class differences in child deprivation in this domain are not very wide. Unlike the survival domain, child deprivation in this domain is the highest in Scheduled Castes but the lowest in Scheduled Tribes. In the development domain, child deprivation is relatively lower than child deprivation in survival and

Table 9.3 Child deprivation in India, 2013–2014

Population	CDI	Domains of deprivation				
		Survival SDI	Growth GDI	Development DDI	Protection PDI	Environment EDI
Total	0.433	0.599	0.426	0.269	0.064	0.455
Residence						
Rural	0.487	0.620	0.436	0.280	0.072	0.616
Urban	0.368	0.550	0.404	0.245	0.045	0.128
Social Class						
SC	0.478	0.617	0.445	0.291	0.075	0.581
ST	0.508	0.630	0.388	0.273	0.064	0.689
OB	0.435	0.592	0.439	0.283	0.060	0.457
OT	0.392	0.582	0.412	0.229	0.058	0.280

Source Author's calculations

growth domains of child well-being. Child deprivation, in this domain, is also the highest in Scheduled Castes but the lowest in Other social classes. Deprivation in the protection domain is the lowest across all domains of child well-being and one possible reason is that RSoC 2013–2014 provides estimates of the proportion of girls aged less than 19 years ever married. A more appropriate indicator of child protection would have been the proportion of girls aged 10–17 years ever married. However, this indicator is not available through RSoC 2013–2014 so that it could not be incorporated in the construction of the child deprivation index. In any case, the deprivation faced by children in this domain is also the highest in Scheduled Castes. Finally, child deprivation in the environment domain is also very high and an important feature of this deprivation, as may be seen from Table 9.3, is very strong residence and social class differences. It is evident from Table 9.3 that child deprivation in India continues to be a major development challenge. Table 9.3 also suggests that residence and social class disparities in child deprivation continue to persist in the country. The RSoC 2013–2014 data also suggest that child deprivation is relatively the highest among Scheduled Tribes children living in the rural areas of the country. Moreover, the challenge of child deprivation is essentially confined to survival, growth and environment domains of child well-being.

9.5 Child Deprivation in States

Estimates of the CDI and deprivation indexes in different domains of child well-being for 29 states of the country are given in Appendix Tables 9.8 through 9.13 for the total population, for rural and urban populations and for different social classes. These tables suggest that the variation across states in child deprivation is quite substantial. It is the lowest in Kerala with a CDI of 0.256 but the highest in Nagaland with a CDI of 0.598. Besides Kerala, Goa is the only other state where child deprivation is very low whereas in Jharkhand, Bihar, Uttar Pradesh, Odisha and Rajasthan, child deprivation is very high as CDI is more than 0.5 in these states. In seven states, child deprivation is low and in 14 states, it is average. There is, however, no state in the country where there is extreme child deprivation, as reflected through CDI (Table 9.4).

Child deprivation has also been found to vary by the social class in all states of the country. There are 10 states in the country where child deprivation in the Scheduled Tribes population may be termed as either high or very high, whereas there is no state where child deprivation in the Scheduled Tribes population is found to be very low as measured through CDI. On the other hand, there are only three states where child deprivation in Other social classes is high but there is no state where child deprivation found to be very high or high in either Other social classes or in Other Backward Classes. Bihar, Jharkhand and Uttar Pradesh are the only three states in the country where the deprivation faced by the children is either high or very high in all the four social classes—Scheduled Castes, Scheduled Tribes, Other Backward Classes and Other social classes.

Table 9.4 Summary measures of variation in child deprivation across states: Total population, residence and social class

Measure	Total	Rural	Urban	Scheduled castes	Scheduled tribes	Other backward classes	Other classes
	Child Deprivation						
Minimum	0.256	0.269	0.249	0.250	0.302	0.248	0.284
1st quartile	0.381	0.411	0.345	0.411	0.430	0.363	0.349
Median	0.439	0.451	0.403	0.451	0.471	0.440	0.421
3rd quartile	0.488	0.555	0.442	0.540	0.590	0.471	0.461
Maximum	0.598	0.632	0.531	0.647	0.645	0.577	0.526
IQR	0.107	0.145	0.097	0.129	0.160	0.108	0.112
	Survival Deprivation						
Minimum	0.367	0.373	0.369	0.352	0.457	0.356	0.411
1st quartile	0.540	0.571	0.496	0.510	0.584	0.495	0.515
Median	0.666	0.674	0.628	0.642	0.683	0.648	0.652
3rd quartile	0.692	0.705	0.682	0.716	0.722	0.688	0.693
Maximum	0.801	0.812	0.769	0.780	0.793	0.770	0.747
IQR	0.152	0.135	0.186	0.206	0.138	0.193	0.178
	Growth Deprivation						
Minimum	0.205	0.222	0.192	0.176	0.207	0.156	0.190
1st quartile	0.319	0.321	0.311	0.330	0.297	0.323	0.352
Median	0.391	0.395	0.395	0.437	0.351	0.427	0.387
3rd quartile	0.453	0.453	0.451	0.477	0.443	0.474	0.454
Maximum	0.583	0.579	0.631	0.616	0.619	0.588	0.579
IQR	0.134	0.132	0.140	0.147	0.146	0.151	0.102
	Development Deprivation						
Minimum	0.074	0.078	0.059	0.056	0.069	0.092	0.052
1st quartile	0.156	0.131	0.191	0.151	0.175	0.146	0.145
Median	0.229	0.250	0.238	0.239	0.279	0.212	0.237
3rd quartile	0.297	0.309	0.292	0.330	0.354	0.330	0.286
Maximum	0.791	0.844	0.629	0.642	0.795	0.852	0.533
IQR	0.141	0.178	0.101	0.179	0.179	0.184	0.141
	Protection Deprivation						
Minimum	0.011	0.006	0.008	0.010	0.010	0.008	0.004
1st quartile	0.035	0.035	0.024	0.037	0.025	0.038	0.020
Median	0.044	0.050	0.037	0.054	0.046	0.049	0.030
3rd quartile	0.067	0.077	0.052	0.096	0.091	0.070	0.043
Maximum	0.342	0.351	0.328	0.329	0.363	0.142	0.174
IQR	0.032	0.042	0.028	0.059	0.066	0.032	0.024

(continued)

Table 9.4 (continued)

Measure	Total	Rural	Urban	Scheduled castes	Scheduled tribes	Other backward classes	Other classes
	Environment Deprivation						
Minimum	0.019	0.034	0.001	0.013	0.012	0.041	0.004
1st quartile	0.139	0.194	0.030	0.196	0.247	0.114	0.112
Median	0.317	0.418	0.074	0.422	0.479	0.155	0.221
3rd quartile	0.434	0.700	0.133	0.658	0.685	0.196	0.322
Maximum	0.777	0.868	0.357	0.896	0.949	0.280	0.605
IQR	0.295	0.506	0.103	0.462	0.438	0.082	0.210

Source Author's calculations

Among different domains of child well-being, deprivation appears to be the highest in the survival domain as the deprivation index in the domain (SDI) is estimated to be more than 0.7 in six states in the total population; nine states in the rural population and ten states in Scheduled Tribes population. Even in the urban population, deprivation faced by children in this domain is found to be extreme in five states. Rajasthan and Uttar Pradesh are the only two states in the country where the deprivation faced by children in this domain is found to be extreme in both rural and urban populations and in all social classes.

The other domain where child deprivation is found to be extreme in many states of the country is the environment domain. There are three states where deprivation in this domain is found to be at its extreme in the total population. In the rural population, the deprivation in this domain is found to be extreme in eight states. Similarly, Scheduled Tribes children in seven states face extreme deprivation in the environment domain. In the urban areas, there is no state where deprivation is extreme in this domain in Other Backward Classes and Other social classes. Deprivation in the development domain is found to be extreme in Nagaland.

Ranking of states in terms of CDI varies by residence and social class (Table 9.5). Kerala is the only state where the rank in CDI in total population, in rural and urban populations and in different social classes is the same. The estimate of CDI in Scheduled Tribes in Kerala is not available through RSoC 2013–2014. The best five states in the context of child deprivation are Kerala, Goa, Sikkim, Tamil Nadu and Andhra Pradesh. The poorest five states are Nagaland, Jharkhand, Bihar, Uttar Pradesh and Odisha. Kerala, Goa, Tamil Nadu and Andhra Pradesh are geographically contiguous. Similarly, Jharkhand, Bihar, Uttar Pradesh and Odisha are geographically contiguous.

The inter-state and within-state rural–urban and social class variation in child deprivation in different domains of child well-being suggests that factors that contribute to child deprivation vary by state and, within each state, by the place of residence and social class in different domains of child well-being. This variation in child deprivation across and within states also suggests that residence and social class

Table 9.5 Ranking of states in terms of CDI, SDI, GDI, DDI, PDI and EDI

State	Population						
	Total	Rural	Urban	SC	ST	OB	OT
	Rank in CDI						
Andhra Pradesh	5	4	4	6	12	5	3
Arunachal Pradesh	13	12	17	na	6	na	22
Assam	17	17	13	9	10	12	19
Bihar	27	26	26	23	19	24	25
Chhattisgarh	23	22	8	16	17	19	8
Delhi	19	18	24	14	na	18	14
Goa	2	3	2	na	1	3	2
Gujarat	9	19	9	5	15	8	9
Haryana	18	16	21	13	11	13	17
Himachal Pradesh	10	7	14	7	8	16	10
Jammu and Kashmir	21	21	27	19	18	21	20
Jharkhand	28	29	25	20	23	23	26
Karnataka	6	6	6	10	4	7	5
Kerala	1	1	1	1	na	1	1
Madhya Pradesh	22	23	10	17	21	17	12
Maharashtra	7	8	7	3	7	6	6
Manipur	12	10	15	na	14	10	11
Meghalaya	20	20	18	na	13	na	23
Mizoram	8	5	11	na	3	na	na
Nagaland	29	28	29	na	20	na	na
Odisha	25	24	19	21	24	20	21
Punjab	15	11	23	11	5	14	18
Rajasthan	24	25	22	18	25	22	24
Sikkim	3	2	5	2	2	2	7
Tamil Nadu	4	13	3	8	na	4	4
Tripura	11	9	12	4	9	9	13
Uttar Pradesh	26	27	28	22	22	25	27
Uttarakhand	14	15	16	15	na	11	16
West Bengal	16	14	20	12	16	15	15
	Rank in SDI						
Andhra Pradesh	2	2	3	2	3	2	2
Arunachal Pradesh	19	16	26	na	13	na	23
Assam	21	21	16	11	16	14	22
Bihar	26	25	27	22	20	21	20

(continued)

Table 9.5 (continued)

State	Population						
	Total	Rural	Urban	SC	ST	OB	OT
Chhattisgarh	9	9	9	10	7	9	7
Delhi	20	20	24	17	na	20	16
Goa	3	5	2	na	1	4	3
Gujarat	8	8	8	6	6	8	9
Haryana	17	17	19	16	14	13	18
Himachal Pradesh	12	11	13	13	15	17	11
Jammu and Kashmir	23	22	20	18	21	25	15
Jharkhand	25	26	21	21	17	22	26
Karnataka	6	6	4	5	4	6	4
Kerala	1	1	1	1	na	1	1
Madhya Pradesh	10	10	10	9	11	10	10
Maharashtra	7	7	6	7	5	7	6
Manipur	24	24	18	na	25	16	14
Meghalaya	15	18	14	na	9	na	25
Mizoram	11	12	12	na	8	na	na
Nagaland	29	29	29	na	24	na	na
Odisha	14	13	22	12	19	12	12
Punjab	16	19	17	14	10	18	19
Rajasthan	27	27	25	19	22	23	24
Sikkim	5	4	7	4	2	3	8
Tamil Nadu	4	3	5	3	na	5	5
Tripura	13	14	11	8	18	11	13
Uttar Pradesh	28	28	28	23	23	24	27
Uttarakhand	22	23	15	20	na	15	21
West Bengal	18	15	23	15	12	19	17
	Rank in GDI						
Andhra Pradesh	14	15	13	9	11	11	7
Arunachal Pradesh	5	5	3	6	5	na	5
Assam	7	7	7	3	4	5	6
Bihar	26	25	25	23	22	23	25
Chhattisgarh	19	21	14	17	19	13	18
Delhi	22	26	22	21	26	22	14
Goa	8	10	8	na	6	8	4
Gujarat	18	18	20	19	17	17	17

(continued)

Table 9.5 (continued)

State	Population						
	Total	Rural	Urban	SC	ST	OB	OT
Haryana	24	22	26	12	23	21	21
Himachal Pradesh	13	14	18	7	10	20	11
Jammu and Kashmir	28	27	29	25	28	25	26
Jharkhand	27	28	17	22	25	24	23
Karnataka	16	16	16	14	8	14	16
Kerala	3	3	6	2	2	3	3
Madhya Pradesh	21	20	24	16	20	15	22
Maharashtra	12	11	12	8	12	9	8
Manipur	10	9	10	na	16	4	9
Meghalaya	11	12	9	18	13	na	12
Mizoram	1	1	1	na	1	na	na
Nagaland	4	4	2	na	7	1	na
Odisha	6	6	5	5	14	7	2
Punjab	25	23	27	20	15	16	24
Rajasthan	23	24	15	13	24	18	19
Sikkim	9	8	11	4	9	6	10
Tamil Nadu	2	2	4	1	3	2	1
Tripura	20	19	21	15	18	19	20
Uttar Pradesh	29	29	28	24	27	26	27
Uttarakhand	15	13	23	11	na	10	13
West Bengal	17	17	19	10	21	12	15
	Rank in DDI						
Andhra Pradesh	10	4	22	8	10	12	9
Arunachal Pradesh	20	20	18	10	14	26	27
Assam	21	19	23	19	20	13	22
Bihar	25	24	26	22	21	23	23
Chhattisgarh	6	8	6	11	6	5	2
Delhi	23	23	24	23	24	25	18
Goa	2	3	2	5	2	1	5
Gujarat	13	14	14	12	11	14	17
Haryana	18	18	16	20	15	18	15
Himachal Pradesh	9	10	21	7	7	11	10
Jammu and Kashmir	19	21	12	17	22	22	16
Jharkhand	24	25	20	21	23	21	25

(continued)

Table 9.5 (continued)

State	Population						
	Total	Rural	Urban	SC	ST	OB	OT
Karnataka	7	7	7	4	8	10	7
Kerala	17	22	12	13	25	19	21
Madhya Pradesh	16	17	9	14	18	15	19
Maharashtra	8	5	11	6	4	7	12
Manipur	4	6	3	na	9	3	3
Meghalaya	28	28	28	24	28	na	26
Mizoram	22	16	25	na	18	8	na
Nagaland	29	29	29	27	29	28	28
Odisha	11	11	15	2	12	9	14
Punjab	14	15	8	16	17	17	11
Rajasthan	26	26	19	25	26	24	20
Sikkim	3	2	4	3	1	4	4
Tamil Nadu	5	9	5	9	5	6	5
Tripura	1	1	1	1	3	2	1
Uttar Pradesh	27	27	27	26	27	27	24
Uttarakhand	12	12	10	18	13	16	8
West Bengal	15	13	17	15	15	20	13
	Rank in PDI						
Andhra Pradesh	25	25	11	18	28	20	8
Arunachal Pradesh	15	13	15	na	18	na	7
Assam	21	20	3	16	23	10	24
Bihar	24	22	22	20	7	17	22
Chhattisgarh	7	7	17	11	10	6	19
Delhi	5	10	6	7	24	2	6
Goa	1	1	5	na	6	1	4
Gujarat	20	22	13	14	11	15	23
Haryana	19	18	20	4	22	21	14
Himachal Pradesh	4	4	3	3	2	5	3
Jammu and Kashmir	3	3	1	1	20	3	1
Jharkhand	23	24	10	19	20	19	10
Karnataka	22	21	22	16	19	22	20
Kerala	12	18	8	24	1	13	13
Madhya Pradesh	12	12	15	7	14	11	15
Maharashtra	17	15	19	12	9	8	25

(continued)

Table 9.5 (continued)

State	Population						
	Total	Rural	Urban	SC	ST	OB	OT
Manipur	10	6	26	na	5	16	17
Meghalaya	15	16	9	15	16	na	1
Mizoram	1	2	2	na	2	na	na
Nagaland	29	29	29	25	29	na	na
Odisha	11	10	25	10	14	12	16
Punjab	6	4	20	4	12	4	9
Rajasthan	26	26	24	21	27	24	21
Sikkim	18	17	18	6	12	14	17
Tamil Nadu	8	8	14	9	4	7	11
Tripura	27	27	27	22	26	18	26
Uttar Pradesh	12	14	7	13	7	9	12
Uttarakhand	8	9	11	2	16	23	5
West Bengal	28	28	28	23	25	25	27
	Rank in EDI						
Andhra Pradesh	21	19	22	18	22	14	15
Arunachal Pradesh	15	15	12	17	12	21	19
Assam	19	17	7	10	14	19	23
Bihar	27	25	27	27	24	28	27
Chhattisgarh	26	27	26	22	25	25	18
Delhi	2	4	5	4	1	5	3
Goa	8	9	17	7	10	9	7
Gujarat	20	21	20	13	21	20	14
Haryana	12	12	16	14	11	13	10
Himachal Pradesh	11	10	13	9	9	12	12
Jammu and Kashmir	16	16	14	19	19	18	17
Jharkhand	28	29	28	24	27	26	26
Karnataka	17	18	18	16	16	15	15
Kerala	1	1	1	3	8	3	2
Madhya Pradesh	25	26	25	23	28	22	22
Maharashtra	18	20	23	12	18	17	20
Manipur	7	7	9	na	2	11	7
Meghalaya	14	13	7	8	13	na	9
Mizoram	4	5	2	na	3	1	na
Nagaland	9	8	11	5	6	8	4

(continued)

Table 9.5 (continued)

State	Population						
	Total	Rural	Urban	SC	ST	OB	OT
Odisha	29	28	29	26	29	27	28
Punjab	6	6	6	6	7	6	5
Rajasthan	24	24	21	21	26	24	21
Sikkim	5	3	4	1	4	4	6
Tamil Nadu	22	22	24	20	17	16	24
Tripura	3	2	3	2	5	2	1
Uttar Pradesh	23	23	19	24	23	23	25
Uttarakhand	10	11	9	15	15	7	11
West Bengal	13	13	15	11	20	10	13

Source Author's calculations

differences in child deprivation have persisted over time and have largely remained uninfluenced by policies and programmes directed towards promoting child well-being.

The variation in CDI and deprivation indexes in different domains of child well-being in India and in its constituent states by the place of residence and by social class calls for a decentralised approach of mitigating child deprivation. This approach may be based on reducing and ultimately eliminating residence and social class inequalities in child deprivation. Residence and social class inequalities in child deprivation may result due to a number of reasons. An important contributor to the observed residence and social class inequalities in child deprivation is the organisational efficiency and administrative capacity of the system that delivers services to children while other reasons may be attributed to a number of factors that are exogenous to the service delivery system. In any case, measurement and analysis of residence and social class inequality in child deprivation may be useful in conceptualising, planning and programming directed towards improving child well-being.

9.6 Child Deprivation Inequality

Inequality among population sub-groups is generally measured in terms of differentials and concentration. Differentials are the most basic. They compare deprivation across groups. The limitation is that differentials consider only between-group differences. Concentration measures, on the other hand, consider all groups and, also, relative size of all groups. Among a number of concentration measures that reflect residence and social class inequality, the coefficient of variation which is a special case of Generalised Entropy measures (Litchfield 1999) is commonly used for three reasons. First, it is based on the variation in deprivation across residence and social class. Second, it is scale independent as it measures variation relative to the average. Third, it can be decomposed into two components—intensity and extensiveness of deprivation. Intensity measures the differential in deprivation while extensiveness measures the proportion of the population to which the observed intensity applies. The square of the coefficient of variation measures the depth of deprivation.

The coefficient of variation is defined as (Firebough 1999):

$$CV = \sqrt{\sum p_i \left(\frac{d_i}{d} - 1\right)^2} \tag{9.7}$$

where d_i is the deprivation index in the residence or social class i, p_i is the proportion of the target population in the class i and d is the deprivation index for all classes combined. Estimates of p_i are, however, not available through RSoC 2013–2014. Therefore, the reduced form of the coefficient of variation was calculated as:

$$CV = \sqrt{\frac{1}{n} \sum \left(\frac{d_i}{d} - 1\right)^2} \tag{9.8}$$

Here n is the number of classes, two—rural and urban—in case of the place of residence and four—social class, Scheduled Castes, Scheduled Tribes, Other Backward Classes and Other classes—in case of social class.

The residence inequality in child deprivation is estimated to be 0.138 for the country as a whole. However, it varies widely across the five domains of child well-being (Table 9.6). In survival, growth and development domains, there is not much difference in child deprivation in rural and urban areas of the country. However, the deprivation gap in the protection domain is quite substantial and this gap is at its extreme in the environment domain. Table 9.6 also suggests that reducing deprivation in the environment and protection domains of child well-being can contribute significantly towards reducing residence inequality in child deprivation in the country.

On the other hand, social class inequality in child deprivation in the country is estimated to be 0.207, which is higher than the residence inequality. Like the residence inequality, social class inequality also varies across different domains of child well-being (Table 9.7). In the environment domain, the social class inequality in deprivation is extremely high. Compared to the environment domain, the social class inequality in child deprivation in all other domains of child well-being is found to be relatively low with the lowest being in the survival domain. The social class inequality in the environment domain of child well-being is almost 10 times the social class inequality in the survival domain of child well-being.

Across the states of the country, both residence and social class inequality in child deprivation varies widely. The residence inequality in child deprivation is found to be highest in Tamil Nadu where CDI in the rural areas is more than 1.5 times the CDI in the urban areas, but lowest in Punjab where CDI in the urban areas is, in fact, marginally higher than that in the rural areas. Chhattisgarh and Madhya Pradesh

Table 9.6 Residence inequality in child deprivation in India and states

Country/State	CDI	SDI	GDI	DDI	PDI	EDI
India	0.138	0.063	0.041	0.069	0.228	0.566
Andhra Pradesh	0.097	0.063	0.019	0.525	0.487	0.533
Arunachal Pradesh	0.024	0.024	0.134	0.052	0.122	0.647
Assam	0.081	0.055	0.114	0.067	0.602	0.662
Bihar	0.124	0.016	0.045	0.048	0.212	0.404
Chhattisgarh	0.231	0.083	0.090	0.168	0.132	0.469
Delhi	0.025	0.005	0.125	0.026	0.707	1.160
Goa	0.083	0.089	0.033	0.127	0.412	0.401
Gujarat	0.158	0.070	0.015	0.034	0.356	0.649
Haryana	0.020	0.014	0.042	0.069	0.112	0.553
Himachal Pradesh	0.009	0.038	0.078	0.541	0.314	0.493
Jammu and Kashmir	0.016	0.022	0.094	0.180	0.343	0.589
Jharkhand	0.173	0.043	0.137	0.156	0.451	0.398
Karnataka	0.103	0.030	0.016	0.197	0.179	0.584
Kerala	0.042	0.012	0.013	0.185	0.374	0.872
Madhya Pradesh	0.196	0.091	0.021	0.162	0.086	0.532
Maharashtra	0.123	0.010	0.018	0.248	0.075	0.563
Manipur	0.044	0.054	0.061	0.024	0.479	0.605
Meghalaya	0.069	0.060	0.056	0.040	0.295	0.664
Mizoram	0.023	0.036	0.072	0.128	0.287	0.898
Nagaland	0.085	0.030	0.148	0.152	0.034	0.565
Odisha	0.154	0.023	0.114	0.163	0.264	0.389
Punjab	0.006	0.021	0.083	0.134	0.504	0.619
Rajasthan	0.139	0.030	0.099	0.196	0.392	0.589
Sikkim	0.051	0.041	0.075	0.179	0.110	0.519
Tamil Nadu	0.259	0.022	0.011	0.054	0.020	0.621
Tripura	0.070	0.090	0.010	0.148	0.161	0.589
Uttar Pradesh	0.089	0.033	0.024	0.115	0.384	0.595
Uttarakhand	0.046	0.071	0.121	0.023	0.083	0.665
West Bengal	0.019	0.007	0.005	0.068	0.114	0.582

Source Author's calculations

Table 9.7 Social class inequality in child deprivation in India and states

Country/State	CDI	SDI	GDI	DDI	PDI	EDI
India	0.207	0.060	0.106	0.123	0.189	0.615
Andhra Pradesh	0.427	0.218	0.164	0.263	1.005	0.937
Arunachal Pradesh	na	na	na	0.653	na	0.852
Assam	0.086	0.082	0.232	0.324	0.615	0.162
Bihar	0.126	0.053	0.116	0.091	0.908	0.235
Chhattisgarh	0.203	0.142	0.068	0.513	0.435	0.378
Delhi	na	na	0.348	0.337	4.655	1.107
Goa	na	na	na	0.544	na	1.024
Gujarat	0.230	0.058	0.129	0.186	0.428	0.804
Haryana	0.065	0.052	0.071	0.179	1.077	0.735
Himachal Pradesh	0.141	0.127	0.286	0.104	0.957	0.429
Jammu and Kashmir	0.266	0.146	0.144	0.356	4.695	1.324
Jharkhand	0.104	0.040	0.043	0.124	0.491	0.261
Karnataka	0.275	0.126	0.287	0.286	0.429	0.956
Kerala	na	na	0.143	0.646	2.946	12.082
Madhya Pradesh	0.292	0.143	0.059	0.255	0.204	0.510
Maharashtra	0.228	0.114	0.073	0.250	0.509	0.629
Manipur	na	na	na	na	na	na
Meghalaya	na	na	na	na	na	na
Mizoram	na	na	na	na	na	na
Nagaland	na	na	na	0.261	na	0.665
Odisha	0.213	0.103	0.224	0.546	0.209	0.292
Punjab	0.043	0.027	0.259	0.398	0.490	0.841
Rajasthan	0.229	0.045	0.120	0.276	0.379	0.554
Sikkim	0.119	0.145	0.071	0.402	0.455	0.928
Tamil Nadu	na	na	0.101	0.266	0.721	0.620
Tripura	0.092	0.145	0.092	0.508	0.501	2.443
Uttar Pradesh	0.119	0.030	0.063	0.142	0.561	0.408
Uttarakhand	na	na	na	0.555	1.738	1.674
West Bengal	0.170	0.020	0.040	0.313	0.367	1.266

Source Author's calculations

are the other states of the country where residence inequality in child deprivation is relatively very high. In contrast, residence inequality in child deprivation is very low in Himachal Pradesh and Jammu and Kashmir.

Residence inequality in child deprivation in different domains of child well-being also varies across states. Residence inequality in the survival domain is the highest in Madhya Pradesh but lowest in Delhi. In the growth domain of child well-being, residence inequality is found to be the highest in Nagaland but lowest in West Bengal, whereas in the development domain, Himachal Pradesh has the highest but Uttarakhand has the lowest residence inequality in child deprivation in this domain. Andhra Pradesh also has very high residence inequality in child deprivation in this domain. Residence inequality in the protection domain is abnormally high in Delhi, a state where the residence inequality in the survival domain is the lowest in the country. Similarly, residence inequality in the protection domain is very low in Tamil Nadu, the state which has the highest residence inequality in child deprivation. In the environment domain also, residence inequality is the highest in Delhi but the lowest in Odisha. Obviously, every state in the country has its own contour of residence inequality in child deprivation. This means that every state has to chart out its own strategy to address the residence inequality in child deprivation.

The social class inequality in child deprivation, on the other hand, is the largest in Andhra Pradesh but lowest in Punjab, although ranking of states in terms of social class inequality in different domains of child well-being is different. Manipur has the largest social class inequality in survival, growth and development domains of child well-being, whereas West Bengal in survival and growth domains and Bihar in the development domain has the lowest social class inequality across Indian states. In the protection domain, social class inequality is the highest in Jammu and Kashmir but the lowest in Madhya Pradesh, whereas in the environment domain, social class inequality is the lowest in Himachal Pradesh but this inequality is at its extreme in Kerala where almost 25% of the Scheduled Tribes households reported practising open defecating compared to less than 1% households belonging to 'Other' social classes.

9.7 Conclusions

This chapter has presented a composite perspective of child deprivation in India and in its constituent states on the basis of the latest data available and the child deprivation index that has been developed for the purpose. The child deprivation perspective incorporates deprivation in survival, growth, development, protection and environment domains of child well-being. An important aspect of the analysis is that it covers, by and large, the entire childhood period, beginning from conception to 18 years of age and thus reflects the whole child perspective.

It is very much evident from the present analysis that child deprivation in India and in its constituent states remains highly pervasive. A substantial proportion of Indian children continue to be devoid of even the basic services and facilities necessary for their survival, growth, development and protection. The analysis also suggests that the proportion of children facing deprivation varies across states and, within states, by residence, social class. Moreover, these differences are quite pervasive in terms of strength and persistence. Such a situation, obviously, cannot be accepted in the context of the best interests of children to which the Government of India is committed at least at the policy level. Observed residence and social class inequality in child deprivation may be attributed to the administrative capacity and organisational efficiency of the efforts directed towards meeting the basic needs of and providing essential facilities to children. It is clear from the analysis that improving the needs effectiveness and increasing the administrative capacity of child well-being efforts may contribute substantially towards addressing child deprivation in India.

The present analysis leads to a number of policy implications for securing and sustaining child well-being. These include:

- Planning and programming for addressing child deprivation must follow an integrated approach. This means that the life cycle approach must be adopted for planning child well-being efforts. This approach should be child-based and not scheme-based, as is the case at present. The Integrated Child Development Scheme launched way back in 1974 was conceptualised on these lines but has currently lost the direction. There is a need to reinvigorate the Scheme to include all pregnant mothers and all children up to 18 years of age. The Scheme must provide continuum of services and facilities to all children.
- There is a need to evolve a system of regularly measuring and monitoring child deprivation through multi-dimensional perspective which can reflect the extent of deprivation faced by children at least up to the district level, if not up to the grassroots level. This system must provide the evidence for planning and programming of child well-being services and for advocating the cause of child well-being, including mobilising community support.
- Although the analysis reveals broad geographic patterns of child deprivation, yet it is obvious that a decentralised approach is required to address issues concerning child well-being. The deprivation faced by children in India is not only residence and social class sensitive but also varies by different domains of child well-being. It is therefore obvious that different states of the country should formulate their own strategic and operational framework to combat the challenge of child deprivation by taking into consideration state-specific child deprivation scenario. Although RSoC 2013–2014 does not provide data that permit estimation of child well-being indicators at district and below district level, yet, given the social, economic, cul-

tural and environmental diversity of the country, it is logical that any approach or effort to address child deprivation must be based upon district-based planning and programming for child well-being in an integrated manner. The analysis also suggests that one of the reasons for the persistence of unacceptable child deprivation scenario in India and in its constituent states is very strong residence and social class inequalities in all domains of child well-being. This observation is important as the analysis also suggests that different forms of deprivation faced by children in different population groups and across states is influenced by a host of cultural, social, economic and environmental factors that operate at the level of the family and the society. The present analysis has not been able to analyse the strength and persistence of these effects which are largely exogenous to any child well-being effort.

- The analysis also does not explore the effect of factors endogenous to child well-being efforts on the deprivation faced by children because of the lack of necessary data. It is, however, well known that the effect of endogenous factors can be minimised, if not completely eliminated, by improving the needs effectiveness and capacity efficiency of the delivery system. There has, however, been little effort to measure needs effectiveness and capacity efficiency of child well-being efforts in India.

Appendix

See Tables 9.8, 9.9, 9.10, 9.11, 9.12 and 9.13.

Table 9.8 Estimates of CDI in India and states by residence and social class: 2013–2014

Country/State	CDI	Residence		Social class			
		Rural	Urban	SC	ST	OB	OT
India	0.433	0.487	0.368	0.478	0.508	0.435	0.392
Andhra Pradesh	0.339	0.383	0.323	0.410	0.462	0.323	0.298
Arunachal Pradesh	0.428	0.442	0.423	na	0.426	na	0.465
Assam	0.444	0.459	0.396	0.413	0.453	0.424	0.456
Bihar	0.581	0.601	0.481	0.647	0.590	0.576	0.521
Chhattisgarh	0.495	0.555	0.345	0.518	0.557	0.471	0.352
Delhi	0.448	0.464	0.448	0.480	na	0.466	0.421
Goa	0.287	0.317	0.272	na	0.302	0.295	0.284
Gujarat	0.394	0.470	0.349	0.404	0.484	0.395	0.376
Haryana	0.444	0.455	0.437	0.467	0.461	0.440	0.441
Himachal Pradesh	0.404	0.406	0.399	0.412	0.431	0.452	0.380
Jammu and Kashmir	0.485	0.496	0.485	0.563	0.579	0.525	0.459
Jharkhand	0.583	0.632	0.449	0.580	0.635	0.572	0.525
Karnataka	0.352	0.395	0.324	0.425	0.412	0.363	0.329
Kerala	0.256	0.269	0.249	0.250	na	0.248	0.284
Madhya Pradesh	0.488	0.560	0.373	0.523	0.615	0.455	0.406
Maharashtra	0.354	0.411	0.330	0.380	0.430	0.353	0.335
Manipur	0.425	0.440	0.403	na	0.481	0.415	0.401
Meghalaya	0.466	0.483	0.424	na	0.471	na	0.465
Mizoram	0.381	0.391	0.375	na	0.380	na	na
Nagaland	0.598	0.623	0.531	na	0.599	na	na
Odisha	0.541	0.575	0.428	0.588	0.636	0.517	0.463
Punjab	0.439	0.440	0.443	0.445	0.422	0.441	0.451
Rajasthan	0.531	0.586	0.442	0.557	0.645	0.531	0.466
Sikkim	0.303	0.298	0.324	0.285	0.311	0.282	0.346
Tamil Nadu	0.334	0.446	0.285	0.413	na	0.313	0.319
Tripura	0.415	0.429	0.376	0.397	0.445	0.401	0.410
Uttar Pradesh	0.574	0.612	0.512	0.618	0.620	0.577	0.526
Uttarakhand	0.434	0.451	0.412	0.492	na	0.419	0.432
West Bengal	0.440	0.451	0.436	0.451	0.514	0.443	0.428

Source Author's calculations

Table 9.9 Estimates of SDI in India and states by residence and social class, 2013–2014

Country/State	SDI	Residence		Social class			
		Rural	Urban	SC	ST	OBC	Others
India	0.599	0.620	0.550	0.617	0.630	0.592	0.582
Andhra Pradesh	0.407	0.393	0.441	0.388	0.493	0.397	0.419
Arunachal Pradesh	0.681	0.675	0.703	0.399	0.683	na	0.712
Assam	0.691	0.700	0.639	0.642	0.701	0.667	0.706
Bihar	0.729	0.731	0.713	0.756	0.747	0.725	0.690
Chhattisgarh	0.569	0.587	0.504	0.632	0.584	0.544	0.488
Delhi	0.687	0.683	0.687	0.714	na	0.693	0.668
Goa	0.435	0.483	0.410	0.417	0.457	0.449	0.426
Gujarat	0.540	0.571	0.496	0.521	0.564	0.535	0.545
Haryana	0.671	0.677	0.659	0.684	0.695	0.648	0.677
Himachal Pradesh	0.638	0.643	0.605	0.655	0.698	0.687	0.599
Jammu and Kashmir	0.696	0.705	0.676	0.717	0.759	0.770	0.664
Jharkhand	0.723	0.735	0.681	0.745	0.707	0.726	0.742
Karnataka	0.476	0.487	0.460	0.498	0.528	0.491	0.448
Kerala	0.367	0.373	0.369	0.352	na	0.356	0.411
Madhya Pradesh	0.607	0.632	0.533	0.623	0.681	0.573	0.554
Maharashtra	0.498	0.503	0.493	0.543	0.531	0.495	0.479
Manipur	0.701	0.726	0.653	na	0.793	0.682	0.652
Meghalaya	0.666	0.679	0.611	0.583	0.664	na	0.728
Mizoram	0.622	0.647	0.602	na	0.621	na	na
Nagaland	0.801	0.812	0.769	0.779	0.793	na	na
Odisha	0.662	0.658	0.682	0.653	0.722	0.637	0.631
Punjab	0.670	0.680	0.653	0.666	0.668	0.688	0.677
Rajasthan	0.732	0.744	0.703	0.732	0.764	0.731	0.718
Sikkim	0.468	0.462	0.494	0.429	0.488	0.431	0.542
Tamil Nadu	0.448	0.441	0.460	0.399	na	0.462	0.466
Tripura	0.647	0.673	0.569	0.602	0.716	0.603	0.633
Uttar Pradesh	0.763	0.773	0.728	0.780	0.768	0.761	0.747
Uttarakhand	0.692	0.720	0.628	0.737	na	0.668	0.696
West Bengal	0.677	0.674	0.683	0.681	0.682	0.688	0.670

Source Author's calculations

Table 9.10 Estimates of GDI in India and states by residence and social class, 2013–2014

Country/State	GDI	Residence		Social class			
		Rural	Urban	SC	ST	OBC	Others
India	0.426	0.436	0.404	0.445	0.388	0.439	0.412
Andhra Pradesh	0.390	0.395	0.381	0.416	0.337	0.402	0.352
Arunachal Pradesh	0.279	0.295	0.229	0.332	0.270	na	0.333
Assam	0.308	0.316	0.259	0.284	0.248	0.285	0.343
Bihar	0.507	0.511	0.475	0.539	0.459	0.496	0.523
Chhattisgarh	0.436	0.451	0.382	0.463	0.430	0.427	0.438
Delhi	0.453	0.533	0.451	0.510	0.590	0.495	0.387
Goa	0.319	0.332	0.311	0.176	0.278	0.325	0.328
Gujarat	0.434	0.441	0.428	0.481	0.407	0.450	0.427
Haryana	0.463	0.453	0.489	0.437	0.470	0.482	0.468
Himachal Pradesh	0.381	0.377	0.422	0.365	0.333	0.477	0.366
Jammu and Kashmir	0.561	0.538	0.631	0.616	0.619	0.561	0.545
Jharkhand	0.517	0.543	0.420	0.521	0.513	0.530	0.483
Karnataka	0.411	0.416	0.403	0.450	0.301	0.427	0.423
Kerala	0.252	0.250	0.256	0.269	0.227	0.244	0.287
Madhya Pradesh	0.443	0.444	0.456	0.461	0.431	0.448	0.470
Maharashtra	0.357	0.352	0.365	0.371	0.339	0.369	0.353
Manipur	0.332	0.327	0.361	0.000	0.367	0.267	0.354
Meghalaya	0.345	0.355	0.320	0.473	0.339	na	0.369
Mizoram	0.205	0.222	0.192	0.000	0.207	na	na
Nagaland	0.268	0.288	0.216	0.216	0.287	0.156	na
Odisha	0.300	0.310	0.253	0.328	0.340	0.323	0.221
Punjab	0.480	0.454	0.530	0.493	0.362	0.449	0.521
Rajasthan	0.455	0.475	0.395	0.446	0.508	0.454	0.439
Sikkim	0.327	0.321	0.361	0.326	0.322	0.312	0.361
Tamil Nadu	0.228	0.231	0.230	0.238	0.232	0.236	0.190
Tripura	0.438	0.441	0.433	0.451	0.412	0.466	0.441
Uttar Pradesh	0.583	0.579	0.603	0.575	0.618	0.588	0.579
Uttarakhand	0.391	0.370	0.455	0.437	na	0.384	0.386
West Bengal	0.423	0.426	0.425	0.431	0.437	0.418	0.416

Source Author's calculations

Table 9.11 Estimates of DDI in India and states by residence and social class, 2013–2014

Country/State	DDI	Residence		Social class			
		Rural	Urban	SC	ST	OBC	Others
India	0.269	0.280	0.245	0.291	0.273	0.283	0.229
Andhra Pradesh	0.176	0.116	0.292	0.152	0.212	0.190	0.159
Arunachal Pradesh	0.282	0.287	0.262	0.183	0.265	0.414	0.442
Assam	0.289	0.285	0.316	0.298	0.311	0.199	0.309
Bihar	0.356	0.353	0.380	0.360	0.325	0.350	0.369
Chhattisgarh	0.139	0.131	0.171	0.193	0.148	0.122	0.054
Delhi	0.323	0.311	0.323	0.383	0.389	0.377	0.260
Goa	0.095	0.110	0.087	0.143	0.076	0.092	0.095
Gujarat	0.227	0.233	0.218	0.194	0.238	0.207	0.253
Haryana	0.266	0.278	0.243	0.308	0.279	0.253	0.241
Himachal Pradesh	0.159	0.146	0.280	0.149	0.158	0.172	0.161
Jammu and Kashmir	0.278	0.303	0.212	0.288	0.354	0.339	0.248
Jharkhand	0.351	0.373	0.277	0.351	0.378	0.327	0.400
Karnataka	0.149	0.130	0.186	0.127	0.175	0.171	0.123
Kerala	0.262	0.309	0.212	0.221	0.426	0.255	0.278
Madhya Pradesh	0.248	0.268	0.195	0.239	0.302	0.217	0.261
Maharashtra	0.156	0.122	0.199	0.145	0.123	0.143	0.180
Manipur	0.122	0.123	0.118	0.000	0.189	0.096	0.063
Meghalaya	0.521	0.528	0.492	0.401	0.538	na	0.402
Mizoram	0.297	0.261	0.337	0.000	0.302	0.147	na
Nagaland	0.791	0.844	0.629	0.642	0.795	0.852	0.533
Odisha	0.194	0.186	0.238	0.114	0.250	0.158	0.233
Punjab	0.228	0.250	0.191	0.276	0.300	0.238	0.177
Rajasthan	0.375	0.406	0.276	0.435	0.441	0.362	0.273
Sikkim	0.101	0.092	0.125	0.117	0.069	0.115	0.075
Tamil Nadu	0.138	0.145	0.130	0.165	0.126	0.135	0.095
Tripura	0.074	0.078	0.059	0.056	0.097	0.095	0.052
Uttar Pradesh	0.479	0.499	0.404	0.519	0.459	0.495	0.382
Uttarakhand	0.202	0.205	0.196	0.296	0.255	0.219	0.152
West Bengal	0.229	0.22	0.249	0.267	0.279	0.258	0.192

Source Author's calculations

Table 9.12 Estimates of PDI in India and states by residence and social class, 2013–2014

Country/State	PDI	Residence		Social class			
		Rural	Urban	SC	ST	OBC	Others
India	0.064	0.072	0.045	0.075	0.064	0.060	0.058
Andhra Pradesh	0.078	0.104	0.031	0.089	0.150	0.078	0.020
Arunachal Pradesh	0.044	0.047	0.037	0.000	0.049	na	0.019
Assam	0.063	0.071	0.010	0.080	0.092	0.044	0.056
Bihar	0.074	0.077	0.052	0.115	0.023	0.068	0.046
Chhattisgarh	0.034	0.032	0.040	0.047	0.033	0.028	0.041
Delhi	0.019	0.038	0.019	0.037	0.105	0.010	0.015
Goa	0.011	0.006	0.015	0.000	0.021	0.010	0.013
Gujarat	0.060	0.077	0.035	0.061	0.035	0.064	0.052
Haryana	0.051	0.055	0.044	0.032	0.090	0.083	0.030
Himachal Pradesh	0.018	0.018	0.010	0.030	0.010	0.026	0.010
Jammu and Kashmir	0.015	0.017	0.008	0.010	0.085	0.017	0
Jharkhand	0.073	0.085	0.028	0.096	0.085	0.070	0.024
Karnataka	0.067	0.075	0.052	0.080	0.057	0.087	0.042
Kerala	0.042	0.055	0.024	0.158	0.000	0.049	0.029
Madhya Pradesh	0.042	0.043	0.037	0.037	0.045	0.045	0.031
Maharashtra	0.047	0.050	0.043	0.050	0.025	0.040	0.058
Manipur	0.039	0.028	0.063	0.000	0.019	0.067	0.040
Meghalaya	0.044	0.051	0.027	0.065	0.047	na	0.004
Mizoram	0.011	0.015	0.009	0.000	0.010	na	na
Nagaland	0.342	0.351	0.328	0.329	0.363	na	na
Odisha	0.041	0.038	0.056	0.042	0.045	0.047	0.032
Punjab	0.027	0.018	0.044	0.032	0.038	0.022	0.023
Rajasthan	0.113	0.131	0.053	0.123	0.129	0.130	0.044
Sikkim	0.049	0.052	0.042	0.033	0.038	0.059	0.040
Tamil Nadu	0.035	0.035	0.036	0.041	0.011	0.038	0.027
Tripura	0.132	0.140	0.103	0.143	0.125	0.069	0.163
Uttar Pradesh	0.042	0.048	0.020	0.054	0.023	0.043	0.028
Uttarakhand	0.035	0.036	0.031	0.026	0.047	0.093	0.014
West Bengal	0.167	0.177	0.142	0.153	0.113	0.142	0.174

Source Author's calculations

Table 9.13 Estimates of EDI in India and states by residence and social class, 2013–2014

Country/State	EDI	Residence		Social class			
		Rural	Urban	SC	ST	OBC	Others
India	0.455	0.616	0.128	0.581	0.689	0.187	0.280
Andhra Pradesh	0.395	0.537	0.133	0.594	0.685	0.169	0.221
Arunachal Pradesh	0.317	0.418	0.045	0.568	0.310	0.171	0.264
Assam	0.375	0.441	0.030	0.338	0.423	0.068	0.369
Bihar	0.748	0.802	0.324	0.896	0.781	0.142	0.570
Chhattisgarh	0.696	0.824	0.252	0.695	0.836	0.155	0.257
Delhi	0.025	0.066	0.024	0.047	0.012	0.082	0.011
Goa	0.139	0.207	0.099	0.181	0.274	0.117	0.112
Gujarat	0.381	0.616	0.122	0.411	0.675	0.217	0.220
Haryana	0.256	0.357	0.083	0.422	0.303	0.081	0.122
Himachal Pradesh	0.218	0.236	0.067	0.251	0.268	0.155	0.178
Jammu and Kashmir	0.331	0.426	0.072	0.643	0.632	0.057	0.231
Jharkhand	0.756	0.868	0.346	0.721	0.904	0.199	0.517
Karnataka	0.332	0.490	0.108	0.551	0.555	0.196	0.221
Kerala	0.019	0.034	0.001	0.042	0.247	0.280	0.006
Madhya Pradesh	0.634	0.803	0.188	0.712	0.905	0.195	0.347
Maharashtra	0.365	0.558	0.148	0.389	0.591	0.219	0.303
Manipur	0.126	0.177	0.031	0.000	0.024	0.067	0.112
Meghalaya	0.306	0.386	0.030	0.210	0.330	0.137	0.114
Mizoram	0.038	0.075	0.007	0.000	0.039	0.144	na
Nagaland	0.140	0.194	0.042	0.070	0.147	0.114	0.018
Odisha	0.777	0.857	0.357	0.889	0.949	0.220	0.605
Punjab	0.097	0.142	0.025	0.149	0.157	0.095	0.060
Rajasthan	0.591	0.749	0.125	0.669	0.877	0.228	0.313
Sikkim	0.044	0.056	0.014	0.013	0.063	0.041	0.063
Tamil Nadu	0.434	0.700	0.161	0.646	0.566	0.265	0.377
Tripura	0.035	0.046	0.008	0.018	0.110	0.163	0.004
Uttar Pradesh	0.577	0.717	0.112	0.721	0.739	0.144	0.392
Uttarakhand	0.216	0.300	0.031	0.441	0.479	0.123	0.170
West Bengal	0.283	0.386	0.074	0.343	0.633	0.189	0.205

Source Author's calculations

References

Chandrashekhar S, Suryanarayana MH (2007) Prevalence of child poverty in India and China. In: Paper presented at the annual meeting of population association of America, New York
Chaurasia AR (2010) Spatial and social class variation in child deprivation in Madhya Pradesh, India. Indian J Hum Dev 4(2):369–390
Citizen's Initiative for the Rights of Children under Age Six (2006) Focus on children under age six. Summary report. New Delhi, Citizen's initiative for the rights of Children under Age Six. Secretariat of the right to food campaign
Dréze J, Khera R, Narayanan S (2007) Early childhood in India: facing the fact. Indian J Hum Dev 1(2):377–388
Firebough G (1999) Empirics of world income inequality. Am J Sociol 104(6):1597–1631
Gordon D, Nandy S, Pantazis S, Pemberton S, Townsend P (2003a) The distribution of child poverty in the developing world. Centre for International Poverty Research, Bristol
Gordon D, Nandy S, Pantazis S, Pemberton S, Townsend P (2003b) Child poverty in the developing world. The Policy Press, Bristol
Government of India (1974) National policy on children. Department of Social Welfare, New Delhi
Government of India (2007) Child protection in the eleventh five year plan (2007–2012). Ministry of Women and Child Development, New Delhi
Government of India (2013) The national policy for children 2013. Ministry of Women and Child Development, New Delhi
Government of India (2015) Rapid survey of children 2013–2014. India fact sheet. Ministry of Women and Child Development, New Delhi
Litchfield JA (1999). Inequality: methods and tools. Text for World Bank's web site on inequality, poverty and socio-economic performance. http://www.worldbank.org/poverty/inequal/index.htm
Minujin A, Delamonica E, Gonzalez E, Davidziuk A (2005) Children living in poverty: a review of child poverty definitions, measurements and policies. In: Paper presented at UNICEF conference on children and poverty: global context, local solutions. New School University, New York
Notten G, de Neubourg C (2011) Monitoring absolute and relative poverty; "Not Enough" is not the same as "Much Less". Rev Income Wealth 57(2):247–269
Nozick R (1974) Anarchy, state and utopia. Blackwell, Oxford
United Nations (1989) United nations convention for the rights of the child. United Nations, New York
Waddington H (2004) Linking economic policy to childhood poverty—a review of the evidence on growth, trade reform and macroeconomic policy. London, Save the Children. The Childhood Poverty and Research Centre. CHIP Report No. 7
White H, Leavy J, Masters A (2002) Comparative perspectives on child poverty: a review of poverty measures. J Hum Dev 4(3):379396

Chapter 10
Female Work Participation in Villages

Abstract This chapter analyses the participation of females in the social and economic production system in more than 578,000 villages in India. The analysis reflects an unsatisfactory female participation in the village-level social and economic production system. The analysis also suggests that the level of female participation in the village economy is influenced by the social class structure and the gender composition of the village population, the level of female education and the level of fertility in addition to the size of the village population. It appears that the economic growth and associated development in India have contributed little to rural women by expanding opportunities for the participation of females in the village-level productive activities.

10.1 Introduction

The ability to earn an income by participating in the social and economic productive activities is widely regarded as an important dimension of women's empowerment, although participation in social and economic productive activities, by itself, may not be regarded as a sufficient condition to ensure an increase in the bargaining power of women and a substantial decision-making role for women within the family and the society (Bardhan 1985; Bennet 1992; Mencher and Saradamoni 1982; Nagaraj 1989; Sinha 2005). In any case, the working status of women, as a major factor in their economic flexibility, has been considered as an indicator of their overall well-being status in the society (Mammen and Paxon 2000). Female participation in productive activities has also been widely recognised as a driver of economic growth and, therefore, women's work participation rates indicate the potential for economic growth (Verick 2014).

Participation of females in the social and economic production system is, however, influenced by a host of economic and social factors that interact in a complex fashion at the level of the household and the society. Global evidence suggests that the key factors influencing women's participation in productive activities include educational attainment, age at the entry into marital union, level of fertility, degree of urbanisation and social norms that determine the role of women in the family and the society. In India, it is argued that the increase in the educational enrolment of young women,

© Springer Nature Singapore Pte Ltd. 2020

A. R. Chaurasia, *Population and Sustainable Development in India*,
https://doi.org/10.1007/978-981-32-9212-3_10

lack of participation opportunities, household income, etc. are some of the factors that influence female participation in productive activities (Chaudhary and Verick 2014; Kapsos et al. 2014). In addition, social norms that decide women's role in the public domain continue to affect outcomes for women. Variation in female work participation rates across the states of India has been found to be influenced by a wide gamut of demographic, social, household and regional factors, although none of these factors are found to be mutually exclusive and can independently explain the variation in female work participation rates across Indian states (Jose 1989; Sinha 2005). Previous studies also suggest that there is no universal set of factors that explains the variation in female work participation rate across the Indian states (Dholakia and Dholakia 1978; Gulati 1975; Nath 1970; Sundaram 1989). This implies that any analysis of participation of females in productive activities within broad socio-economic and cultural perspective, such as the country as a whole or state or even district and sub-district, does not appear to be the right approach of understanding the dynamics of female work participation. Rather, it appears to be more appropriate to analyse female work participation at the grassroots level and to identify the factors and conditions that influence this participation. There is, however, little attempt in this direction. Analysis of female work participation in India has generally been carried out at national and state levels (Chandrasekhar and Ghosh 2011; Chaudhary 2011; Rangarajan et al. 2011; Sanghi et al. 2015; Vinoj 2013). Village-level analyses of female work participation are rare in India. Sinha (2005) has carried out an analysis of female participation in productive activities in villages of four districts of West Bengal, whereas Rogers (2012) has used village-level surveys to analyse labour force participation in 36 villages of Bihar. These studies have highlighted the variation in female work participation across villages because of a number of village-level factors. There has, however, been little attempt to analyse variation in female work participation across the villages of the country. There has also been little attempt to identify village-level factors that contribute to variation in female work participation across villages.

The above considerations constitute the rationale for the present analysis. We measure, in this chapter, female work participation in more than 578 thousand villages of the country and explore distinct patterns of female work participation across villages on the basis of a set of village-level characteristics following the data mining approach. Data mining is the process of discovering patterns in big data so as to extract information and transform the information into an understandable structure from the perspective of policy-level analysis and programme-level interventions (Hastie et al. 2008).

Participation in social and economic productive activities, essentially, has two dimensions—the dimension of engagement and the dimension of duration of engagement. The dimension of engagement reflects the extensiveness of participation while the dimension of duration of engagement indicates the intensity of participation. The two dimensions, in combination, determine the level of participation (Heckman 1993). There is, however, no study in India, to the best of our knowledge, that takes into account both the dimensions of participation in studying female work participation. Female work participation has generally been analysed in terms of female labour force participation rate or female work participation rate which is defined as the proportion of females engaged in any social and economic productive activity. Labour force participation rate and work participation rate are actually indicators of the extensiveness of participation. They do not take into account the intensity of participation in social and economic productive activities.

This chapter develops an index of participation in social and economic productive activities that takes into account bothdimensions of participation in productive activities—the dimension of extensiveness and the dimension of intensiveness—and use this index to measure and analyse female work participation across village of India. The analysis leads to the classification or segmentation of villages of the country into mutually exclusive groups or clusters of villages having distinct village-level characteristics that are associated with different levels of female participation in village-level social and economic production system. The analysis reveals that the female participation at the village level is influenced by the defining characteristics of the village. The analysis also reveals that the composition of the female workforce at the village level varies across different groups or clusters of villages having distinct village-level characteristics.

The chapter is organised as follows: The next section describes the index of participation used to measure the level of participation in social and economic production system. Section three describes the data used in the present analysis. The analysis is based on the primary census abstract (PCA) of the 2011 population census, which is the only source of data related to participation of the people in productive activities at the village level in India. Section four of the chapter presents and discusses results of the analysis, while the last section of the chapter summarises the main findings of the analysis and discusses their implications in the context of improving the participation of females in the productive activities at the village level.

10.2 The Index of Participation

The level of participation in productive activities in a village can be measured in terms of the total duration of engagement of the people of the village in productive activities within a reference period which may be a day or a month or a year or even a lifetime (Blundell et al. 2011). Assuming the reference period to be a year, the level of participation in the productive activities is then the product of the number of people engaged in productive activities during the year and the average number of days of engagement per person per year. If the number of people engaged in the productive activities in a village is L and the average number of days of engagement per person per year in the village is A, then the total duration of engagement, D, of the people of the village in productive activities in a year is:

$$D = L * A \tag{10.1}$$

The above conceptualisation suggests that if E is the proportion of the population of the village who is engaged in productive activities in a year and I is the ratio of the average number of days of engagement of a person to the total number of days in the year, then the index of participation in productive activities, P, may be defined as:

$$P = E * I \tag{10.2}$$

It is obvious that the higher the value of the index P, the higher the level of participation in productive activities. Here, the index E measures the extensiveness of participation in productive activities and is nothing but the conventional work participation rate. The higher is the number of people engaged in productive activities, the higher is the value of E. On the other hand, the index I measures the intensity of participation in productive activities. The higher the average duration of engagement of a person in productive activities in a year, the higher the value of I. It is also obvious by definition that both E and I vary between 0 and 1 and, therefore, it is also obvious that the index of participation, P, defined as the product of the measure of the extensiveness and the measure of the intensiveness of participation measured in terms of the average duration of participation per person in a year also varies between 0 and 1.

The index of participation in productive activities, P, defined in the above manner, may be calculated for specific occupational categories also. If Ej is the extensiveness of participation and Ij is the intensity of participation in the productive activity j, then the activity-specific index of participation, Pj, may be defined as:

$$P_j = E_j * I_j \tag{10.3}$$

The index of participation for all productive activities may then be defined as:

$$P = \sum_j P_j = \sum_j E_j * I_j \tag{10.4}$$

The change in the index P over time may then be decomposed, following Kittagawa (1955), in the following manner:

$$\nabla P = P_2 - P_1 = \sum_j E_{2j} * I_{2j} - \sum_j E_{1j} * I_{1j}$$

or

$$\nabla P = \sum_j (E_{2j} - E_{1j}) * \bar{I}_j + \sum_j \bar{E}_j + (I_{2j} - I_{1j}) = \sum_j \partial E_j + \sum_j \partial I_j = \partial E + \partial I \tag{10.5}$$

where $\bar{I}_j = (I_{2j} + I_{1j})/2$, etc.

We use the above analytical formulation to measure and analyse the participation of females in social and economic productive activities at the village level. For every village of the country, we estimate the index of participation for different work categories by calculating the category-specific index of extensiveness of participation and category-specific index of intensiveness of participation of females. The category-specific index of extensive of participation and the index of intensiveness of participation have then been combined to obtain the index of female participation in the social and economic productive activities in the village. It is obvious that female participation in the village social and economic production system depends upon the opportunities of participation available in the local level social and economic production system.

10.3 Data

The analysis is based on the primary census abstract (PCA) of the 2011 population census which provides data on the work status of the population for every village of the country. Work, in India's 2011 population census, is defined as participation in any economically productive activity with or without compensation, wages or profit (Government of India 2011). Participation may be physical and/or mental. Work, according to the 2011 population census, involves not only the actual work but also supervision and direction. Part time help or unpaid work on farm, family enterprise or in any other economic activity has also been classified as work. People who are engaged in cultivation or milk production even solely for domestic consumption are also classified as workers. Workers enumerated at the population census are classified into three categories on the basis of the duration they worked during the year prior to the census—workers who worked for at least 6 months; workers who worked for 3–6 months; and workers who worked for less than 3 months. During the year prior to the census, workers are further classified into one of the four occupational categories—cultivators; agricultural labourers; household industry workers; and other workers. A worker is classified as cultivator if she or he is engaged in cultivation of his or her own land or land owned by other individuals or institutions including government for payment in money, kind or share. Cultivation includes effective supervision or direction in cultivation. A person who has given out her/his land to another person or persons or institution(s) for cultivation for money, kind or share of crop and who does not even supervise or direct cultivation of land, is not treated as cultivator. Similarly, a person working on another person's land for wages in cash or kind or a combination of both is not treated as cultivator. Cultivation involves ploughing, sowing, harvesting and production of cereals and millet crops such as wheat, paddy, jowar, bajra, ragi and other crops such as sugarcane, tobacco, groundnuts, tapioca and pulses, raw jute and kindred fibre crop, cotton, cinchona and other medicinal plants, fruit growing, vegetable growing or keeping orchards or groves, and so on. Cultivation does not include the following plantation crops—tea, coffee, rubber, coconut and betel-nuts (areca).

On the other hand, a person who works on another person's land for wages in money or kind or share is classified as agricultural labourer. She or he has no risk in the cultivation, but merely works on another person's land for wages. An agricultural labourer has no right of lease or contract on the land on which she/he works. Similarly, a worker is classified as household industry worker if she or he is engaged in a household industry, which is an industry conducted by one or more members of the household at home or within the village in rural areas and only within the precincts of the house where the individual lives in urban areas. The household industry is not run on the scale of a registered factory. In the urban areas, even if household members run an industry by themselves but at a place away from the precincts of their home, it is not considered as a household industry. Household industry relates to production, processing, servicing, repairing or making and selling (but not merely selling) of goods. It does not include professions such as a pleader, doctor, musician, dancer, waterman, astrologer, dhobi, barber, etc. or merely trade or business. Lastly, all workers who are not classified as either cultivators or agricultural labourers or household industry workers are classified as other workers. They include government servants, municipal employees, teachers, factory workers, plantation workers, workers engaged in trade; commerce; business; transport; banking; mining; construction; political or social work, priests, entertainment, artists, etc.

The concept of the village adopted in the Indian population census is different from the commonly used concept of the village as a human settlement, which is larger than a hamlet but smaller than a town. A hamlet has a tiny population less than 100 (Doxiadis 1968). During the census, the urban areas are identified first on the basis of clearly laid down definition of a standard urban area and the population living in these urban areas is classified as the urban population. Population not residing in the urban areas, on the other hand, is classified as the rural population which is then organised into administrative areas following the administrative boundaries of revenue villages and these administrative areas are termed as villages in the census parlance. A village, defined during the population census in India, is, therefore, an administrative unit with well-defined administrative boundaries and the population of the village so defined is the number of persons enumerated within the administrative boundaries of the village at the time of the census. This approach of defining a village during the census means that a village may have one or more than one human settlements within its administrative boundaries or it may have no human settlement at all in which case, the population of the village is zero. If there are more than one human settlements within the administrative boundaries of a village then PCA 2011 provides data of all human settlements combined and not data pertaining to separate human settlements within the village. This approach of defining a village pays no attention to the permanent or temporary nature of human settlements as the enumeration is carried out on the *de facto* basis and not on the *de-jure* basis. There is always a possibility that human settlements within the boundaries of a village are permanent or temporary or both.

According to PCA 2011, there were 640,949 villages in the country at the time of 2011 population census. There were 43,330 or 6.8% villages where no population was enumerated at the time of 2011 population census so that the number of inhabited villages in 2011 was 597,619 or 93.2% of the total villages identified. The population of these villages varied from just 1 to 66,062 persons. In 18,841 or 2.9% villages, the number of households was less than 10. Moreover, there were 23 inhabited villages where there was no female population. We have excluded all these villages from the present analysis so that the present analysis is restricted to 578,755 villages or the lowest level administrative units in the rural areas of the country as identified at the time of the 2011 population census. These villages account for 90% of the villages of the country listed at the time of 2011 population census and the population living in these villages account for 99.9% of the total rural population of the country—population living in those areas which have not been classified as urban at the time of the 2011 population census.

10.4 Methods

Estimating the index of participation, P, for the villages of the country requires estimation of the index of extensiveness of participation, E, and the index of intensiveness of participation in the village. We have measured the index of extensiveness of participation in terms of the ratio of the workers to the total population in the village. A more appropriate indicator to measure the extensiveness of participation would have been the ratio of workers aged 15–59 years to the population aged 15–59 years. However, PCA 2011 does not provide data pertaining to population and workers in the villages by age. Although the ratio of workers to total population is a crude measure of the extensiveness of participation, yet, it gives an idea about how many people in the village are engaged in productive activities—the higher the ratio, the higher the extensiveness of participation.

On the other hand, the index of intensiveness of participation, I, is estimated as the ratio of average annual days of work per person per year divided by the potential number of days of work per person per year. For estimating the average annual number of days of work per person, we assumed that frequencies of each work interval are centred at the mid value of the interval. Thus, we assumed that workers who worked for at least 6 months during the year prior to the census actually worked for 270 days in a year—270 is approximately the mid value of the interval 180–365 days. Similarly, we assumed that workers who worked for 3–6 months in the year prior to the census actually worked for 135 days in a year and workers who worked for less than 3 months actually worked for 45 days in the year. Under these assumptions, the average annual days of work in a year per person (*adw*) is calculated as:

$$adw = \frac{(270 * W_M + 135 * W_G + 45 * W_L)}{W} \tag{10.6}$$

where W_M is the number of workers who worked for at least 6 months in the year, W_G is the number of workers who worked for 3–6 months in the year and W_L is the number of workers who worked for less than 3 months in the year. Once, *adw* is estimated, the index of intensiveness of participation, I, is calculated as:

$$I = \frac{adw}{270} \tag{10.7}$$

Here, it is assumed that the total number of potential days available for work for a person is 270 days.

It is logical to assume that the index of participation, P, is influenced by a host of village-specific characteristics. These characteristics include but are not limited to (1) population of the village; (2) level of education in the village; (3) level of fertility in the village; (4) gender composition of the village population; and (5) social class structure of the village population. We have measured the level of education in the village in terms of the effective literacy rate, which is defined as the proportion of population aged 7 years and above who can read and write with understanding. Similarly, the level of fertility has been surrogated by the ratio of children aged 0–6 years to females aged 7 years and above. This ratio is very similar to the familiar child–woman ratio, which is widely used as a crude indicator of fertility (Shryock and Siegel 1980). Obviously, the higherthe ratio, the higher the level of fertility.

In order to analyse how village characteristics influence the participation in productive activities in the village, we have followed the classification modelling approach. This approach classifies or segments villages into different groups or cluster of villages in such a manner that the within-group or within-cluster homogeneity with respect to the index of participation is the highest. The classification modelling approach is different from the regression-based approach that is commonly used for analysing marginal effects of the defining characteristics of the village on the index of participation (Chaurasia 2012). Unlike the regression-based approach, there is no restriction or limitation on the structure of the independent variables or the defining characteristics of the village which are used as explanatory variables in the classification modelling exercise. In the most general terms, the classification or the segmentation emanating from the classification modelling exercise is based on a set of *if–then* logical conditions that permit splitting or classifying or segmenting villages into mutually exclusive groups or classes of villages.

We have used the Decision Tree procedure to create a tree-based classification or segmentation model which classifies villages into groups or clusters of villages by partitioning villages into smaller groups or classes so that villages within a group are as homogenous as possible. The procedure can be used for many purposes and we use the procedure here for the segmentation of villages (IBM Corporation 2012). Within-group or cluster homogeneity is measured in terms of entropy. The lower is the entropy, the higher is the homogeneity. There are many tree-growing methods which can be used for classification modelling. We have used the classification and regression tree (CRT) method (Breiman et al. 1984), which is a non-parametric recursive partitioning method. CRT splits villages into groups or segments that are as homogenous as possible with respect to the dependent variable—index of participation. A terminal node in which all villages have the same value for the dependent variable is called a homogeneous or 'pure' node. The extent to which a node does not represent a homogenous subset of villages is an indication of impurity. There are different impurity measures available. In case of continuous or scale variables, impurity is measured in terms of least-squared deviation which is computed as the within-node variance adjusted for frequency weights or influence variables as the case may be (IBM Corporation 2012).The tree-growing process is continued until either the pure node is reached or the prescribed stopping criterion is met (Ambalavanan et al. 2006; Lemon et al. 2003).

CART has a number of advantages as an exploratory data analysis procedure for the purpose of classification or segmentation, which is the primary objective of the present analysis. It makes no assumption about the distribution of the dependent variable or the independent variables used in the analysis. Moreover, explanatory or the independent variables used in CART can be a mix of categorical, interval and continuous or scale variables. Another advantage of CART is that results of the analysis are not at all affected by the quality of data such as presence of outliers, collinearity among explanatory variables, heteroscedasticity or distributional error structures that normally affect parametric procedures.

10.5 Female Participation in Villages

On the basis of the data available through the 2011 population census, we have estimated the index of female participation in productive activities, P_f, for 578,755 villages included in the present analysis. P_f is found to vary widely across villages of the country (Table 10.1). There are 2954 villages in the country where P_f is estimated

Table 10.1 Distribution of villages by index of female participation, P_f, in different occupational categories in India, 2011

P_f		All	Cultivation	Agricultural labour	Household industry	Others
Very low	<0.15	37.7	77.4	71.3	98.8	95.7
Low	0.15–0.30	26.7	12.3	19.8	0.9	3.3
Average	0.30–0.45	20.1	6.2	6.7	0.2	0.8
High	0.45–0.60	13.2	3.3	1.9	0.1	0.2
Very high	≥0.60	3.3	0.8	0.3	0.0	0.0
Weighted mean		0.224	0.068	0.104	0.011	0.041
Unweighted mean		0.250	0.096	0.107	0.009	0.038
Standard deviation		0.178	0.140	0.126	0.033	0.058
Median		0.228	0.026	0.056	0.001	0.019
IQR		0.284	0.123	0.164	0.005	0.033
Skewness		0.540	1.948	1.486	8.076	4.268

Source Author's calculations

to be zero, whereas there are 20 villages where P_f is estimated to be 1 at the time of 2011 population census. The distribution of villages by P_f has been found to be positively skewed. which means that in majority of the villages, P_f is less than the average. In nearly two-thirds of the villages of the country, P_f is estimated to be less than 0.30 and in almost 38% villages, it is estimated to be less than 0.15. In contrast, there are only about 3% villages where P_f is estimated to be 0.60 and more. The weighted average of P_f across 578,755 villages is estimated to be 0.224 whereas the unweighted average is estimated to be 0.250. The weighted average takes into account the size of the population of the village which varies widely across villages. The fact that the weighted average of P_f is lower than its unweighted average implies that P_f is relatively lower than the average in more villages than in villages where P_f is higher than the average.

Table 10.1 also summarises the distribution of villages in terms of the index of female participation by work categories—cultivation (P_{fc}), agricultural labour (P_{fa}), household industry work (P_{fh}) and in other productive activities (P_{fo}). The table indicates that the index of female participation in household industry, P_{fh}, is very low. There are only about 1% villages where P_{fh} is at least 0.15. Similarly, there are less than 5% villages where P_{fo} is at least 0.15. In contrast, the index of female participation in cultivation, P_{fc} and in agricultural labour, P_{fa} is estimated to be 0.15 and more in about 23% and 29% villages of the country, respectively. At the same time, there are 37 villages in the country where the index of female participation in household-level productive activities, P_{fh}, is estimated to be very high—0.60 and more. Similarly, in 272 villages of the country, the index of female participation in productive activities other than cultivation, agricultural labour and household industrial activity, P_{fh}, is estimated to be very high.

The index of female participation in productive activities may also be viewed as a reflection of the participation opportunities available for females in the village level social and economic production system and it can be assumed that the widerthe opportunities for participation of females available in the village-level social and economic production system, the higher the female participation in the productive activities. In this context, Table 10.1 suggests that the opportunities for female participation in the village-level productive activities appear to be seriously limited in nearly all the villages of the country and whatever opportunities are available they are virtually confined to the agriculture sector, particularly, agricultural labour. There appears to be very limited opportunities of productive participation of females in sectors other than agriculture at the village level and this appears to be an important reason behind low to very low participation of females in village social and economic production system.

The index of female participation, P_f, is determined by the index of extensiveness of female participation, E_f, and the index of intensiveness of female participation, I_f. Table 10.2 presents the weighted average of E_f and I_f by occupational category across villages of the country. The E_f and I_f for all occupational categories combined is the sum of occupational category specific E_f and I_f. It may be seen from the table that the variation in E_f across occupational categories is more marked than the variation in I_f . Moreover, E_f is the highest in the occupational category agricultural labour but I_f in this occupational category is very low. On the other hand, both E_f and I_f are the lowest in the occupational category household industry.

The difference in the index of female participation P_f between two occupational categories is due to the difference in both the index of extensiveness E_f and the index of intensiveness I_f. This difference can be decomposed into the difference attributed to the difference in the extensiveness and the difference in the intensiveness of participation in the two categories. For example, almost all the difference between P_{fa} and P_{fh} is attributed to the difference in the extensiveness of participation. The difference attributed to the difference in the intensiveness of participation as agricultural labour and as household industry worker is, at best, marginal. This observation suggests that the very low index of female participation in the household industry in the villages of the country is largely due to very limited opportunities of female participation in household-level productive activities. This means that if female participation in the household industry sector is to be increased, then it is imperative that opportunities for female participation in the household-level productive activities at the village level must be expanded. The intensiveness of female participation is not a major issue in improving female participation in the household productive activities at the village level. The female participation in other sectors of the village social and economic production system is also, at present, very low.

10.6 Classification of Villages

Data available through the 2011 population census clearly shows that the female participation in productive activities varies widely across the 578,755 villages of India. It is also well known that Indian villages vary widely in terms of selected defining characteristics of the village population. This means that variation in the index of female participation in village-level productive activities needs to be analysed in terms of the variation in a set of defining characteristics of the village population. We have carried out classification modelling exercise to examine this association. The classification modelling exercise essentially involved building the classification tree through CRT methodology. The dependent variable in the classification modelling exercise is the index of female participation P_f which is estimated for 578,755 villages of the country. The independent or classification variables used for the purpose

Table 10.2 Weighted average of the index of extensiveness and the index of intensiveness in villages of India by occupation category

Index	All	Cultivation	Agricultural labour	Household industry	Others
E_f	0.300	0.087	0.145	0.015	0.053
I_f	0.748	0.790	0.716	0.711	0.775
P_f	0.224	0.069	0.104	0.011	0.041

Source Author's calculations

of modelling include: (1) total population of the village; (2) population sex ratio measured in terms of the ratio of the number of males aged 7 years and above to the number of females aged 7 years and above in the village; (3) proportion of Scheduled Tribes females to total females in the villages; (4) effective female literacy rate in the village measured in terms of the proportion of females aged 7 years and above in the village who can read and write with understanding; and (5) level of fertility in the village measured in terms of the ratio of the population aged 0–6 years to females aged 7 years and above.

Results of classification modelling exercise are presented in Table 10.3 and the classification tree is depicted in Fig. 10.1. The first split of 578,755 villages of the country included in the present analysis is on the proportion of Scheduled Tribes females to total females in the village—307,100 (53.1%) villages where there was virtually no Scheduled Tribes population at the 2011 population census (Node 1) and 271,655 (46.9%) villages where Scheduled Tribes were present (Node 2). The unweighted average of P_f across villages of Node 1 is substantially lower than that in villages of Node 2. Next, villages of Node 1 are further split on the level of fertility in the 127,103 (22.0%) villages, where the proportion of children aged 0–6 years to females aged 7 years and above is less than or equal to 0.316 (Node 3) and 179,997 (31.1%) villages where this ratio is more than 0.316 (Node 4). The unweighted average of the P_f across villages of Node 3 is 0.242 compared to 0.178 across villages of Node 4. On the other hand, villages of Node 2 are further split into 181,084 (31.1%) villages where the proportion of Scheduled Tribes females to total females in the village ranged between 0.003 and 0.638 (Node 5) and 90,571 (15.6) villages where the proportion of Scheduled Tribes females to total females in the village is more than 0.638 (Node 6). The unweighted average of the P_f across villages of Node 5 is 0.281 compared to 0.339 across villages of Node 6.

At the third level of classification, villages of Node 3 are further split into 46,637 (8.1%) villages where effective female literacy is less than or equal to 0.628 (Node 7) and 80,466 (13.9%) villages where the effective female literacy is more than 0.628 (Node 8). The unweighted average of P_f across villages of Node 7 is 0.295 compared to 0.212 across villages of Node 8. On the other hand, villages of Node 4 are split further on the population of the villages into 75,467 (13.0%) villages having population less than or equal to 842 (Node 9) and village having population more than 842 (Node 10). The unweighted average of P_f across villages of Node 9 is 0.208 compared to 0.157 across villages of Node 10. Similarly, villages of Node 5 are split into 58,384 (10.1%) villages having effective female literacy rate less than or equal to 51.3% (Node 11) and 122,700 (21.2%) villages having effective female literacy rate of more than 51.3% (Node 12). The unweighted average of P_f across villages of Node 11 is 0.308 compared to 0.268 across villages of Node 12. There is no further splitting of villages of Node 6 so that Node 6 is a terminal node.

At the last level of classification, villages of Node 7 are further split on the ratio of children 0–6 years to females aged 7 years and above into 16,837 (2.9%) villages where this ratio is less than or equal to 0.243 (Node 13) and 29,800 (5.1%) villages where this ratio ranges between 0.243 and 0.316 (Node 14). The unweighted average of P_f across villages of Node 13 is 0.352 compared to 0.283 across villages of Node

Table 10.3 Female work participation rate by defining characteristics of villages

Node/Cluster number	Proportion of scheduled tribes females	Male/Female ratio in population aged at least 7 years	Ratio of children 0–6 years to females aged at least 7 years	Female effective literacy rate	Population	Distribution of the index of female participation (P_f) within the cluster			Number of villages in the cluster
						Unweighted average	Standard deviation	Skewness	
6	>0.638					0.339	0.168	0.198	90571
9	≤0.003		>0.316		≤842	0.208	0.181	0.771	75467
10	≤0.003		>0.316		>842	0.157	0.132	1.171	104530
12	0.003–0.638			>0.513		0.268	0.165	0.423	122700
13	≤0.003		≤0.243	≤0.628		0.352	0.194	−0.077	16837
14	≤0.003		0.243–0.316	≤0.628		0.263	0.184	0.420	29800
15	≤0.003	≤1.001	≤0.316	>0.628		0.255	0.187	0.572	31170
16	≤0.003	>1.001	≤0.316	>0.628		0.185	0.175	1.214	49296
17	0.003–0.638		≤0.316	≤0.513		0.382	0.162	−0.269	15384
18	0.003–0.638		>0.316	≤0.513		0.281	0.156	0.274	43000
All						0.250	0.178		578755

Source Author's calculations

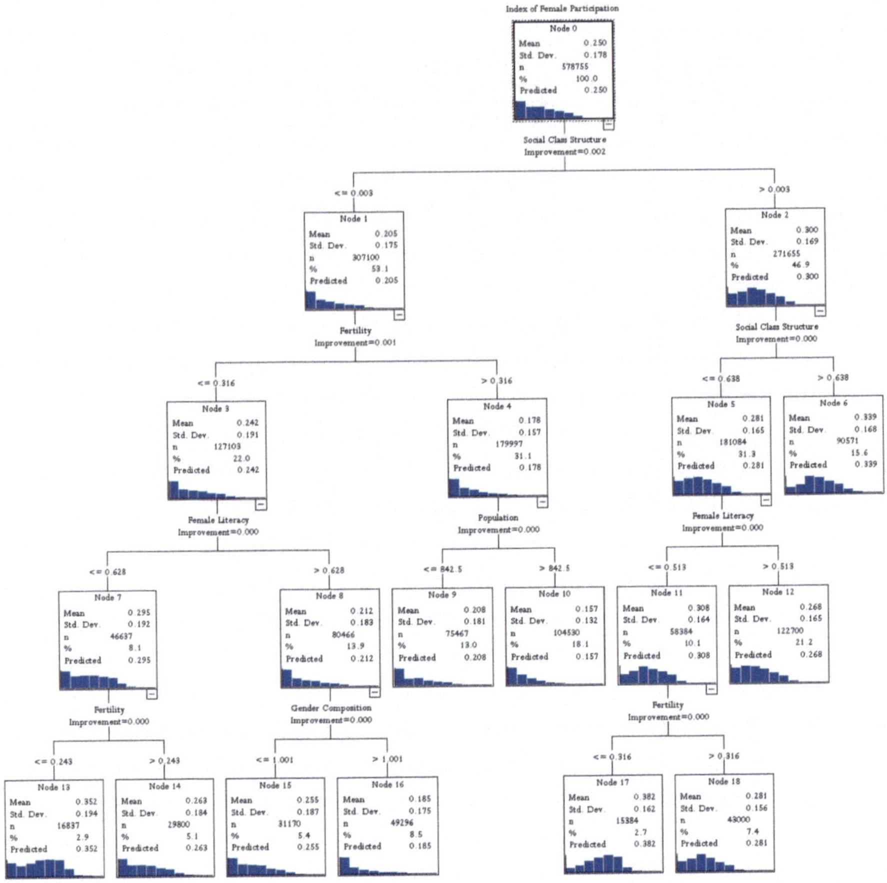

Fig. 10.1 Classification of villages by their defining characteristics and level of female work participation rate

14. On the other hand, villages of Node 8 are split on the ratio of males aged 7 years and above to females 7 years and above into 31,170 (5.4%) villages with this ratio less than or equal to 1.001 (Node 15) and villages with this ratio greater than 1.001 (Node 16). The average of P_f across villages of Node 15 is 0.255 compared to 0.185 across villages of Node 16. There is no further split of villages of Node 9 and Node 10, whereas villages of Node 11 are split further on the ratio of children 0–6 years to females 7 years and above into 15,384 (2.7%) villages with this ratio less than or equal to 0.316 (Node 17) and 43,000 (7.4%) villages having this ratio more than 0.316 (Node 18). The unweighted average of P_f across villages of Node 17 is 0.382 compared to 0.281 across villages of Node 18. Finally, there is no further split of villages of Node 12.

As can be seen from Fig. 10.1, the classification exercise yields 10 terminal nodes which means that 578,755 villages of the country can be grouped into 10 mutually exclusive nodes or clusters of villages in the context of the index of female participation in productive activities at the village level. The defining characteristics of villages of different clusters are shown in Table 10.3. The unweighted average of P_f of villages of cluster 10 is the lowest but that of cluster 17 is the highest amongst the 10 clusters. Other clusters where the unweighted average of P_f is relatively high are cluster 13 and cluster 6, whereas the unweighted average of P_f is relatively low in cluster 16 and cluster 9. It may also be observed from Fig. 10.1 and Table 10.3 that within-cluster distribution of villages in terms of P_f is generally positively skewed with the exception of cluster 13 and cluster 17 where the within-cluster distribution of villages in terms of P_f. A positive skewness means that in majority of the villages within the cluster, P_f is less than the average P_f for the cluster but there are villages where P_f is higher than the cluster average. The degree of skewness in within-cluster distribution of villages by P_f, however, varies across clusters, being very high in clusters 16 and 12 but very low in clusters 6 and 18. In clusters 13 and 17, however, P_f is higher than the cluster average in majority of the villages in the cluster but there are villages in these clusters where P_f is smaller than the unweighted cluster average.

The patterns of female participation in productive activities in different clusters of villages are presented in Table 10.4. The variation in the index of female participation in productive activities, P_f, across clusters is due to both variation in the index of extensiveness of participation, E_f, and variation in the index of intensiveness of participation, I_f. There are only three clusters where more than 40% of females were found to be engaged in productive activities at the time of 2011 population census, whereas in two clusters, this proportion was less than 25%. On the other hand, there are only two clusters where the intensity of participation was more than 80% whereas in two clusters, it was less than 70%.

More specifically, the very high index of female participation, P_f, in cluster 17 is due to the highest index of the extensiveness of participation E_f and the second highest index of intensiveness of participation I_f. Similarly, P_f is the second highest in cluster 13 because E_f is the third highest in the cluster but I_f is the highest. In contrast, P_f is the lowest in cluster 10 because E_f is the lowest and I_f is the second lowest. Similarly, P_f is the second lowest in cluster 16 primarily because E_f is the lowest in this cluster, although I_f ranks sixth in this cluster. In cluster 9, P_f is the third lowest primarily because I_f is the lowest in this cluster, although E_f ranks seventh across clusters. Table 10.4 also reveals that below average E_f in clusters 12 and 15 is associated with the above average I_f in these clusters. There is, however, no cluster in which both E_f and I_f are the highest or the lowest across clusters.

Table 10.4 also presents patterns of female participation across clusters by different occupational categories. The index of female participation as cultivator, P_{fc} is the highest in cluster 6, not in cluster 17. However, the index of female participation as agricultural labourer and household industry workers is the highest in cluster 17. On the other hand, the index of female participation as other occupations is the highest in cluster 15. In contrast, female participation as cultivator, agricultural labour and other occupations is the lowest in cluster 10 but participation in household productive activities is the lowest in cluster 6. It is evident from Table 10.4 that these variations

Table 10.4 Effectiveness and intensiveness of female participation in different clusters

Index	Node/Cluster										
	6	9	10	12	13	14	15	16	17	18	All
Scheduled Tribes	*>0.638*	*≤0.003*	*≤0.003*	*0.003–0.638*	*≤0.003*	*≤0.003*	*≤0.003*	*≤0.003*	*0.003–0.638*	*0.003–0.638*	*All*
Sex ratio (7+)	*All*	*All*	*All*	*All*	*All*	*All*	*≤1.001*	*>1.001*	*All*	*All*	*All*
Children (0–6)	*All*	*>0.316*	*>0.316*	*All*	*≤0.243*	*0.243–0.316*	*≤0.316*	*≤0.316*	*≤0.316*	*>0.316*	*All*
Literacy	*All*	*All*	*All*	*>0.513*	*≤0.628*	*≤0.628*	*>0.628*	*>0.628*	*≤0.513*	*0.513*	*All*
Population	*All*	*≤842*	*>842*	*All*	*All*	*All*	*All*	*All*	*All*	*All*	*All*
	All workers										
E_f	0.470	0.290	0.213	0.318	0.439	0.332	0.286	0.221	0.487	0.370	0.300
I_f	0.712	0.686	0.690	0.786	0.848	0.770	0.786	0.765	0.839	0.706	0.748
P_f	0.335	0.199	0.147	0.250	0.372	0.256	0.225	0.169	0.409	0.261	0.224
	Cultivators										
E_{fc}	0.176	0.114	0.058	0.080	0.123	0.101	0.084	0.061	0.111	0.118	0.087
I_{fc}	0.797	0.732	0.734	0.825	0.887	0.806	0.754	0.784	0.905	0.768	0.790
P_{fc}	0.140	0.083	0.043	0.066	0.109	0.081	0.063	0.048	0.100	0.091	0.069
	Agricultural labourers										
E_{fa}	0.235	0.119	0.098	0.162	0.235	0.163	0.101	0.083	0.299	0.191	0.145
I_{fa}	0.653	0.634	0.647	0.763	0.831	0.749	0.785	0.740	0.815	0.669	0.716
P_{fa}	0.153	0.075	0.063	0.124	0.195	0.122	0.079	0.061	0.244	0.128	0.104

(continued)

Table 10.4 (continued)

Index	Node/Cluster										
	6	9	10	12	13	14	15	16	17	18	All
	Household industry workers										
E_{fh}	0.011	0.013	0.016	0.015	0.018	0.013	0.015	0.016	0.027	0.013	0.015
I_{fh}	0.619	0.650	0.679	0.735	0.840	0.730	0.791	0.698	0.860	0.651	0.711
P_{fh}	0.007	0.008	0.011	0.011	0.015	0.009	0.012	0.011	0.023	0.008	0.011
	Other workers										
E_{fo}	0.048	0.044	0.041	0.062	0.062	0.056	0.086	0.062	0.052	0.048	0.053
I_{fo}	0.711	0.722	0.735	0.810	0.840	0.776	0.817	0.796	0.825	0.716	0.775
P_{fo}	0.034	0.032	0.030	0.050	0.052	0.043	0.070	0.049	0.043	0.034	0.041

Source Author's calculations

in the index of female participation can be traced in variation in both the index of extensiveness and the index of intensiveness of participation in different occupations across clusters identified through classification modelling. It may also be observed from Table 10.4 that very high index of extensiveness of female participation, P_f, in clusters 6, 13 and 17 is largely because of very high index of extensiveness of female participation as agricultural labourers. At the same time, very low index of extensiveness in female participation, E_f, in cluster 10 is primarily because of very low index of extensiveness of female participation as cultivators, P_{fc}, and agricultural labourers, P_{fa}. Similarly, very high index of intensiveness of female participation, I_f, in clusters 13 and 17 is largely because of very high index of intensiveness of female participation in all the four occupational categories. By comparison, the index of intensiveness of female participation in all the four occupational categories is generally very low in cluster 10. These observations suggest that opportunities for female participation in village-level productive activities vary widely across the 10 clusters of villages identified through classification modelling exercise.

The difference in the index of female participation, P_f between two clusters may be decomposed into the difference resulting from the difference in the index of intensiveness of female participation, E_f, and the difference resulting from the difference in the index of intensiveness of female participation, I_f. Results of this decomposition analysis are presented in Table 10.5. For example, the difference between the index of female participation in cluster 17, the cluster with highest P_f and cluster 10, the cluster with lowest P_f is 0.261 in absolute terms. More than 78% of this difference is attributed to the difference in the index of extensiveness of female participation, E_f, between the two clusters while the difference in the index of intensiveness of female participation, I_f, accounts for about 22% of the difference in P_f. More specifically, the difference in the index of extensiveness of female participation as agricultural labour, E_{fa} alone accounts for more than 56% of the difference in the index of female participation, P_f, between cluster 17 and cluster 10. At the same time, the difference in the intensiveness of female participation as agricultural workers between the two clusters accounts for almost 13% of the total difference. In other words, the very high participation of females in villages of cluster 17 is primarily due to very high index of female participation as agricultural labourers in cluster 17 as compared to cluster 10. In cluster 17, not only the index of extensiveness of female participation as agricultural labourer is higher than that in cluster 10 but also the index of intensiveness of female participation as agricultural labourer is also high.

Table 10.5 also shows that the difference in the index of female participation in villages of different clusters from the index of female participation in villages of cluster 10 may be attributed to the difference in female participation in different occupational categories. For example, the difference in the index of female participation in villages of cluster 15 and cluster 16 from that of cluster 10 may be attributed largely to the difference in the index of female participation as other workers whereas this difference in clusters 6 and 9 may be attributed largely to the difference in the index of female participation as cultivators. In the case of other clusters, this difference is attributed largely to the difference in the index of female participation as agricultural labourers. Even in villages of cluster 6, the difference in the index of female par-

Table 10.5 Decomposition of the difference of the index of female participation, P_f, in a cluster from the index of female participation in cluster 10, the cluster with the lowest index of female participation

Cluster	∇P_f	Difference attributed to (%)													
		∇E_f	∇I_f	Cultivation			Agricultural labour			Household industry work			Others		
				∇P_{fc}	∇E_{fc}	∇I_{fc}	∇P_{fa}	∇E_{fa}	∇I_{fa}	∇P_{fh}	∇E_{fh}	∇I_{fh}	∇P_{fo}	∇P_{fo}	∇P_{fo}
16	0.022	32.1	67.9	21.2	7.6	13.6	−8.0	−46.6	38.6	0.8	−0.6	1.4	86.0	71.7	14.3
9	0.052	105.2	−5.3	78.8	79.3	−0.5	22.9	25.8	−2.9	−4.4	−3.6	−0.8	2.6	3.7	−1.1
15	0.077	71.4	28.6	26.0	24.2	1.8	20.9	3.1	17.8	1.6	−0.6	2.2	51.5	44.7	6.8
12	0.103	74.7	25.3	22.5	16.4	6.1	58.1	43.5	14.6	0.2	−0.6	0.8	19.2	15.4	3.8
14	0.109	80.0	20.1	35.3	30.0	5.3	53.6	41.4	12.2	−1.3	−2.0	0.7	12.5	10.6	1.9
18	0.114	95.9	4.2	42.0	39.4	2.6	56.3	53.6	2.7	−2.0	−1.6	−0.4	3.8	4.5	−0.7
6	0.187	96.6	3.4	51.9	48.0	3.9	48.1	47.6	0.5	−2.1	−1.7	−0.4	2.1	2.7	−0.6
13	0.225	76.7	23.4	29.6	23.4	6.2	58.6	45.0	13.6	2.0	0.8	1.2	9.9	7.5	2.4
17	0.261	78.7	21.4	22.3	16.7	5.6	69.1	56.3	12.8	3.8	2.4	1.4	4.9	3.3	1.6

Source Author's calculations

Remarks $\nabla P_f = \nabla E_f + \nabla I_f$

$$= \nabla P_{fc} + \nabla P_{fa} + \nabla P_{fh} + \nabla P_{fo}$$

$$= \nabla E_{fc} + \nabla E_{fa} + \nabla E_{fh} + \nabla E_{fo} + \nabla I_{fc} + \nabla I_{fa} + \nabla I_{fh} + \nabla I_{fo}$$

$\nabla P_{fc} = \nabla E_{fc} + \nabla I_{fc}$, etc

ticipation as agricultural labourer is quite substantial. Interestingly, the difference in the index of female participation as household industry workers is not found to be substantial in any cluster.

It may also be seen from Table 10.5 that the contribution of the difference in the index of extensiveness of female participation is substantially higher than that of the index of the intensiveness of female participation in all but one cluster. One reason for the relatively low contribution of the difference in the index of intensiveness of female participation to the index of female participation may be the way the index of intensiveness of female participation has been calculated because of data constraints. It has been assumed in calculating the index of intensiveness in female participation that the engagement in a productive activity, on average, is either for 45 days in a year or 135 days or 270 days. The reason for this hard assumption is that PCA 2011 provides data about the number of females who worked for 1–90 days in a year; 90–180 days in a year and 180–365 days in a year. For the calculation of the index of intensiveness of female participation, the middle value or the simple arithmetic mean of each interval has been taken as representing the interval. During the 2011 population census, data pertaining to the actual number of days of engagement in productive activities during the year prior to the census was not collected from the working population. Rather, every worker enumerated at the census was asked whether she or he worked for 6 months or more or for 3–6 months or for less than 3 months during the year prior to the census. The estimates of average duration of engagement in productive activities are therefore derived from the limited set of data. As such, variation in the index of intensiveness of female participation across villages is substantially limited.

10.7 Regional Patterns

Table 10.6 presents the distribution of villages of different states/Union Territories of the country across different clusters identified through the data mining exercise. For the country as a whole, almost 68% of the villages are found to be confined to clusters 6, 9, 10 and 12 but the distribution of villages across different clusters is found to be different in different states/Union Territories. For example, more than 95% villages in Kerala are confined to only two clusters of villages—clusters 12 and 15. In contrast, more than 95% villages in Delhi and more than 87% villages in Haryana are found to be confined to clusters 10 and 16. Similarly, almost 80% villages in Bihar and almost 75% villages in Uttar Pradesh are found to be confined to clusters 9 and 10 only. In the north-eastern states of the country, virtually all villages are found to be confined to cluster 6 only. In other states/Union Territories, however, villages are found to be distributed across more than two clusters. For example, villages in Jammu and Kashmir are almost equally distributed across clusters 9, 10, 12 and 18. The same appears to be the case in Rajasthan where majority of villages are distributed almost equally across five clusters—clusters 6, 9, 10, 12 and 18. In Andhra Pradesh, more

Table 10.6 Distribution of villages in states/Union Territories by node/Cluster

Country/State/Union Territory	Node/Cluster										All
	6	9	10	12	13	14	15	16	17	18	
Jammu and Kashmir	9.196	16.1	17.72	22.08	1.685	5.392	3.434	6.291	0.658	17.45	100
Himachal Pradesh	3.993	15.29	0.888	16.05	2.841	2.727	35.15	22.08	0.195	0.781	100
Punjab	0	8.232	7.516	0	4.755	16.45	12.2	50.85	0	0	100
Chandigarh	0	0	100	0	0	0	0	0	0	0	100
Uttarakhand	1.925	28.15	5.495	7.791	8.192	7.52	30.46	9.48	0.129	0.859	100
Haryana	0	9.285	40.02	0	1.69	16.18	1.476	31.35	0	0	100
Delhi	0	3.884	45.63	0	0	0	0.971	49.51	0	0	100
Rajasthan	15.29	19.92	15.75	11.07	2.076	6.848	0.581	1.65	4.088	22.72	100
Uttar Pradesh	0.255	25.42	49.46	5.896	1.268	7.185	2.152	5.867	0.09	2.412	100
Bihar	1.405	23.46	55.99	7.671	0.346	1.082	0.385	1.337	0.114	8.21	100
Sikkim	19.42	0.485	0	78.16	0.243	0	0.243	0	0.971	0.485	100
Arunachal Pradesh	85.82	2.51	0.591	5.685	0.07	0.148	0.07	0.148	0.763	4.184	100
Nagaland	94.66	0.07	0	4.037	0	0	0.07	0	0.288	0.865	100
Manipur	79.55	3.638	3.719	4.77	0.283	0.97	4.77	1.536	0.323	0.445	100
Mizoram	98.56	0	0	0.432	0	0	0	0	0	1.007	100
Tripura	49.13	0.348	2.897	39.05	0	0	1.043	7.068	0	0.464	100
Meghalaya	95.38	0.996	0.67	1.829	0.03	0.07	0.05	0.1	0.03	0.849	100
Assam	22.85	13.69	18.09	23.22	0.531	1.999	4.995	9.488	0.633	4.505	100
West Bengal	7.821	5.86	9.828	36.26	1.784	5.184	3.625	21.32	3.454	4.858	100
Jharkhand	36.07	18.03	10.03	13.53	0.413	1.401	0.356	0.985	1.221	17.97	100

(continued)

Table 10.6 (continued)

Country/State/Union Territory	Node/Cluster										All
	6	9	10	12	13	14	15	16	17	18	
Odisha	30.07	4.726	1.105	25.26	1.234	2.709	7.702	14.27	3.811	9.12	100
Chhattisgarh	39.18	3.7	3.659	38.54	0.227	1.173	0.569	0.532	1.897	10.52	100
Madhya Pradesh	24.64	13.36	8.746	30.61	0.759	3.418	0.608	1.729	1.586	14.55	100
Gujarat	24.91	6.482	17.47	23.83	1.647	5.529	3.097	12.31	0.519	4.209	100
Daman and Diu	15.79	0	15.79	63.16	0	0	0	5.263	0	0	100
Dadra & Nagar Haveli	93.85	0	0	6.154	0	0	0	0	0	0	100
Maharashtra	13.55	3.823	5.079	53.35	2.424	2.843	7.248	9.47	0.644	1.575	100
Andhra Pradesh	17.76	1.949	2.335	19.74	12.82	11.57	1.656	1.64	25.38	5.137	100
Karnataka	2.362	3.672	4.053	43.06	11.45	5.963	7.989	7.552	5.989	7.914	100
Goa	7.692	0.641	0	49.68	0	0.321	19.55	22.12	0	0	100
Lakshadweep	100	0	0	0	0	0	0	0	0	0	100
Kerala	0.1	0	2.753	48.67	0	0	47.3	0.983	0.197	0	100
Tamil Nadu	1.953	1.672	1.946	19.21	18.2	14.12	24.97	16.29	1.271	0.368	100
Puducherry	0	0	3.333	0	0	4.444	85.56	6.667	0	0	100
AN Islands	19.38	17.19	1.563	16.88	2.188	0.625	6.875	35.31	0	0	100
India	15.65	13.04	18.06	21.2	2.912	5.149	5.386	8.517	2.658	7.429	100

Source Author's calculations

than 87% villages are distributed across five clusters of villages 6, 12, 13, 14 and 17. Similarly, almost 93% villages in Tamil Nadu are distributed across five clusters.

Viewed differently, more than 40% villages of cluster 6 are located in—Jharkhand, Chhattisgarh and Madhya Pradesh whereas almost 55% villages of cluster 9 are located in Rajasthan, Uttar Pradesh and Bihar, with Uttar Pradesh alone accounting for almost one-third villages of this cluster. Nearly two-thirdvillages of cluster 10 are located in Uttar Pradesh and Bihar with Uttar Pradesh accounting for almost 45% villages of this cluster. Similarly, almost 54% of villages of cluster 13 are located in Andhra Pradesh, Karnataka and Tamil Nadu and around 43% villages of cluster 14 are located in Rajasthan, Uttar Pradesh and Andhra Pradesh. More than half of the villages of cluster 15 are located in Himachal Pradesh, Uttarakhand, Odisha and Tamil Nadu whereas around 53% villages of cluster 16 are located in only four states—Punjab, Uttar Pradesh, West Bengal and Odisha. More than three-fourths villages of cluster 17, on the other hand, are located in only four states—Rajasthan, Odisha, Andhra Pradesh and Karnataka—with Andhra Pradesh, alone, accounting for more than 42% villages of this cluster. Finally, more than half of the villages of cluster 18 are located in only three states—Rajasthan, Jharkhand and Madhya Pradesh.

10.8 Conclusions

The present analysis has attempted to measure female participation in social and economic productive activities in more than 578 thousand villages of India on the basis of an index of female participation that takes into account both the extensiveness and the intensiveness of participation in productive activities. The analysis indicates that the level of female participation in the village economy in India remains far from satisfactory. There are very few villages in the country where female participation in productive activities may be termed as satisfactory. More importantly, there is only nominal participation of females in household-level productive activities in all but a few villages and one probable reason may be that there are very limited opportunities of household-level productive activities in the village economy in India. The level of female participation in the villages of the country is largely dependent upon the level of female participation as agricultural labour.

The analysis also suggests that the level of female participation in the village economy is influenced by the social class structure and the gender composition of the village population, the level of female education and the level of fertility in the village in addition to the size of the village population. The level of female participation is found to be associated positively with the proportion of Scheduled Tribes population in the village but negatively with the level of fertility, level of female education and the size of the village population. The level of female participation has also been found to be relatively low in those villages where the gender balance is not in favour of females. When these five defining characteristics of the village are taken into consideration, more than 578 thousand villages of the country can be grouped into

10 mutually exclusive clusters, each having significantly different level of female participation. The 10 clusters of villages identified in the present analysis also have distinct regional patterns with villages of some clusters confined largely to one or two states/Union Territories only.

A revealing finding of the present analysis is that the level of female participation in village economy is negatively associated with the level of female education in the village as measured by the female literacy rate. This negative association contradicts the widely held argument that female education increases opportunities for the participation of women in productive activities. It appears that opportunities of participation of educated females in the economy of the villages of the country are very limited. If the index of extensiveness of female participation is any indication, then work, in the villages of the country, is largely confined to the agriculture sector and that too in the form of agriculture labour. It is argued that females prefer women-centric work which can be discharged from the household or within the household premises in the nature of self-employment (Sanghi et al. 2015). However, such opportunities of participation appear largely absent in the villages of India according to the data available through the 2011 population census. The village economy in India remains largely traditional with little sign of transition despite substantial expansion and growth of country's economy in the recent past. It appears that the growth and expansion of the economy of the country have largely been irrelevant to Indian villages where almost 70% of India's population lives. This also means that the economic growth and associated development in India appears to have contributed little to rural women by expanding opportunities for the participation of females in the village-level productive activities.

References

Ambalavanan N, Baibergenova A, Carlo WA, Saigal S, Schmidt B, Thorpe KE (2006) Early prediction of poor outcome in extremely low birth weight infants by classification tree analysis. J Pediatr 148(4):438.e1–444.e1

Bardhan K (1985) Women's work, welfare and status: forces of tradition and change in India. Econ Polit Wkly 22(50, 51, 52)

Bennet L (1992) Women, poverty and productivity in India. Seminar Paper No. 43, Economic Development Institute. Washington DC, The World Bank

Blundell R, Bozio A, Laroque G (2011) Extensive and intensive margins of labour supply: Working hours in the US, UK and France. Discussion Paper No. 6051. Bonn, The Institute for the Study of Labour

Breiman L, Friedman J, Olshen RA, Stone CJ (1984) Classification and Regression Trees. Wadsworth

Chandrasekhar CP, Ghosh J (2011) Latest employment trends from NSSO. Business Line, July 12

Chaudhary R, Verick S (2014) Female labour force participation in India and beyond. New Delhi, International Labour Organization. ILO DWT for South Asia and ILO Country Office for India. ILO Asia-Pacific Working Paper Series

Chaudhary S (2011) Employment data in India. What does latest data show? Econ Polit Wkly 46(32)

Chaurasia AR (2012) Contraceptive use in India: a data mining approach. Int J Popul Res 2014(821436): 11 pages. http://dx.doi.org/10.1155/2014/821436

Dholakia BH, Dholakia RH (1978) Inter-state variation in female labour force participation rates in India. Indian J Labor Econ 20(4)

Doxiadis CA (1968) Ekistics: an introduction to the science of human settlements. Oxford University Press, New York

Government of India (2011) Census of India 2011. Meta Data. New Delhi, Office of the Registrar General and Census Commissioner

Gulati L (1975) Female work participation: a study of inter-state differences. Econ Polit Wkly 10(1, 2)

Heckman J (1993) What has been learned about labor supply in the past twenty years? Am Econ Rev 83(2):116–121

Hastie T, Tibshirani R, Friedman J (2008) The elements of statistical learning. data mining, inference and prediction. Springer

IBM Corporation (2012) IBM SPSS Decision Trees 21. IBM Corporation

Jose AV (1989) Limited options: women workers in rural India. Asian Regional Team for Employment Promotion and World Employment Programme, International Labour Office

Kapsos S, Silberman A, Bourmpoula E (2014) Why is female labour force participation declining so sharply in India? Geneva, International Labour Office. ILO Research Paper No. 10

Kitagawa EM (1955) Components of a difference between two rates. J Am Stat Assoc 50(272):1168–1194

Lemon SC, Roy J, Clark MA, Friedmann PD, Rakowski W (2003) Classification and regression tree analysis in public health: methodological review and comparison with logistic regression. Ann Behav Med 26(3):172–181

Mammen K, Paxon C (2000) Women's work and economic development. J Econ Perspect 14(4):141–164

Mencher J, Saradamoni P (1982) Muddy feet, dirty hands: rice production and female agricultural labour. Econ Polit Wkly 17(52)

Nagaraj K (1989) Female workers in rural Tamil Nadu: a preliminary study. In: Jose AV (ed) Limited options: women workers in rural India. Asian Regional Team for Employment Promotion and World Employment Programme, International Labour Office

Nath K (1970) Female work participation and economic development. Econ Polit Wkly 15(31)

Rangarajan C, PI Kaul, Seema (2011) Where is the missing labour force? Econ Polit Wkly 46(39):68–72

Rogers J (2012) Labour force participation in rural Bihar: a thirty-year perspective based on village surveys. New Delhi, Institute for Human Development. WP04/2012

Sanghi S, Sirja A, Vijay SS (2015) Decline in rural female labour force participation in India. A relook into causes. Vikalpa. J Decis Makers 40(3):255–268

Shryock SH, Siegel JS, Larmon A (1980) The methods and materials of demography, vol I. United States Bureau of Census

Sinha S (2005) Female work participation rates in rural West Bengal: a village level analysis. Indian J Labour Econ 48(3):563–577

Sundaram K (1989) Inter-state variations in participation rates of women in India: an analysis. In Jose AV (ed) Limited options: women workers in rural India. Asian Regional Team for Employment Promotion and World Employment Programme, International Labour Office

Verick S (2014) Women's labour force participation rate in India. Why is it so low? New Delhi, ILO DWT for South Asia and Country Office for India

Vinoj A (2013) Missing labour force or 'defeminization' of labour force in India? Thiruvananthapuram, Centre for Development Studies. CDSWP 452

Chapter 11
Population Effects of the Environment

Abstract This chapter analyses the impact of population growth on sustainable development processes in India. The analysis shows that technology advancement had and is likely to have only a limited impact on the environmental concerns in India. The decrease in the primary energy use and the resulting decrease in CO_2 emission attributed to the improvement in the technology used in the social and economic production system and has been found to be too small to compensate for the increase in the primary energy use and CO_2 emission as a result of population growth and the increase in affluence. As such, the case of accelerating population transition in the context of sustainable development in India becomes even stronger.

11.1 Introduction

It is well known that pervasive poverty, environmental degradation and hazards, use of natural resources, and economic and social development are closely linked to population growth and distribution. It is also recognised that unsustainable patterns of consumption and production are depleting natural resources and causing environmental degradation, while reinforcing social inequity and poverty. The challenge of the sustainability of development processes is to meet the needs of the present generation and improve their quality of life in ways that do not compromise the ability of future generations to meet their own needs. It is therefore important to recognise the inter-relationship between population and the three dimensions of sustainable development identified in the 2030 Sustainable Development Agenda - economic growth, social inclusion and environmental protection (United Nations 2015a, b). It is imperative in the context of sustainable development that the population dimension is explicitly recognised in all development strategies directed towards economic growth, social inclusion and environmental protection. The medium and the most likely variant of 2017 population projections prepared by the United Nations suggests that the world population is most likely to increase from an estimated 7383 million in 2015 to 8551 million in 2030, or an increase of 1168 million during 15 years period (United Nations 2017). Almost 97% of this increase (1131 million) will be confined to the so-called developing or less developed countries of the world, although

© Springer Nature Singapore Pte Ltd. 2020
A. R. Chaurasia, *Population and Sustainable Development in India*,
https://doi.org/10.1007/978-981-32-9212-3_11

there will be substantial variation in the rate of population growth across countries. Along with the very substantial, increase in population, there will also be significant changes in the age structure of the population of different countries because of the transition in fertility and mortality and movement of the population. All these changes in the population stock will undoubtedly have implications to the demand for natural resources and the impact on the environment in terms of the wastes generated out of the natural resources use. The change in the population stock will also influence economic growth processes and prospects of human development, including poverty eradication and unequal distribution of resources across countries and within countries. These concerns were recognised at the International Conference on Population and Development at Cairo in 1994. The Conference emphasised the need of integrating population factors in development planning and programming and recommended that population issues should be explicitly taken into consideration in economic and human development strategies planning, decision-making and resource allocation at all levels and in all regions—and argued that integration of population factors in development planning will contribute to both sustainable development and poverty alleviation and will lead to slowing down the population growth and improving the quality of life (United Nations 1994). However, population dimension of sustainable development has surprisingly been ignored in the 2030 Sustainable Development Agenda of the United Nations.

The population dimension of sustainable development is not just about the implications of the increase in the size of the population but also about the implications of changes in the population age structure in the form of demographic dividend, population ageing and mobility—urbanisation and migration. These dimensions of population are central to any development process directed towards meeting the needs of the people and, therefore, have direct bearings on the realisation of the sustainable development goals and associated targets as identified in the United Nations 2030 Sustainable Development Agenda.

It is in the above context, we analyse, in this chapter, how population growth impacts upon sustainable development processes in India. The United Nations has estimated that India's population in 2015 was around 1309 million (United Nations 2017). The medium variant of the projected population of the country prepared by the United Nations suggests that India's population is most likely to increase to 1513 million by the year 2030 (United Nations 2017). This means that during the 15 years period between 2015 and 2030, India's population is most likely to increase by around 204 million. It is obvious that this increase in the size of the population of the country will have important implications to economic and human development processes of the country and will have substantial environmental impact which may be characterised in terms of primary energy use and associated carbon dioxide (CO_2) emission. An understanding of the contribution of the increase in the population to the increase in the primary energy use and the increase in CO_2 emission may reflect the population dimensions of sustainable development in India.

The chapter is organised as follows. The next section of the chapter describes the methodology adopted for analysing the impact of population growth on primary energy use and associated CO_2 emission. We use the most commonly used IPAT framework to analyse the contribution of the increase in population to the increase in the primary energy use and CO_2 emission. The third section of the chapter describes the data source used in the analysis. Results of the analysis are presented in section four of the chapter while the fifth section discusses policy implications of the findings of the analysis in the context of sustainable development. The last section of the chapter summarises main conclusions of the analysis and emphasises the need of integrating population factors in sustainable development processes in India.

11.2 Methodology

The present analysis employs the well-known IPAT framework that assumes that the environmental impact is the combined effect of population, affluence and technology. This framework was first proposed in the early 1970s (Ehrlich 1968; Commoner 1971) and is still used widely. The framework assumes that all anthropogenic changes in the environment are primarily due to (i) population (ii) economic activity, (iii) technology, (iv) political and economic institutions and (v) attitudes and beliefs (Stern et al. 1992). The framework is simple, systematic and robust. It is simple because it incorporates key anthropogenic driving forces with parsimony. It is systematic because it specifies the mathematical relationship between the driving forces and their impact; and it is robust because it is applicable to a wide variety of environmental impacts (Dietz and Rosa 1997).

The IPAT framework requires measurement of the environmental impact and its three components—population, affluence and technology. The environmental impact has been measured in terms of primary energy use and associated CO_2 emission. The primary energy use reflects the extent of the use of natural resources, whereas CO_2 emission reflects the technology involved in transforming natural resources into usable form. On the other hand, the most straightforward measure of population in the context of its environmental impact is the size of the population. Other things being equal, the larger is the population, the larger the demand and use of natural resources and the larger the waste generated out of the natural resources use. On the other hand, affluence represents the average consumption per person—the higher the average consumption, the higher the use and hence the demand for natural resources and hence the higher the waste generated out of natural resources use. A common proxy for measuring consumption or the natural resources use is the gross domestic product GDP per capita. Finally, the technology component of IPAT framework may be measured in terms of primary energy intensity of the gross domestic product, which is defined as the primary energy used in producing one unit of the gross domestic product. Many factors influence the energy intensity of the gross domestic product. These include, among others, general standards of living and weather conditions, structure of the social and economic production system and the efficiency of primary energy use by the social and economic production system. The energy intensity of the gross domestic product also reflects the state of the technology. When the energy intensity of the gross domestic product is low then, other things being equal, the state of technology may be termed as advanced as compared to when the energy intensity of the gross domestic product is high. It may also be noticed that the energy intensity of the gross domestic product also reflects the use of natural resources.

If G denotes the total gross domestic product or the total output of the social and economic production system and P denotes the size of the population, then the IPAT framework may be represented in terms of the total primary energy use (E) in the following manner:

$$E = P * \frac{G}{P} * \frac{E}{G} = P * A * R \tag{11.1}$$

Here A denotes the real gross domestic product per capita while R denotes the primary energy intensity of the gross domestic product. In this formulation, environmental impact is measured in terms of total primary energy use.

Equation (11.1) may be extended further in terms of total CO_2 emission to what is known as the Kaya identity (1990):

$$C = P * \frac{G}{P} * \frac{E}{G} * \frac{C}{E} = P * A * R * W \tag{11.2}$$

Here W denotes the carbon intensity of energy use, which means CO_2 emitted in using one unit of primary energy. CO_2 is one of the greenhouse gases. Increased concentration of these gases in the atmosphere is attributed to disruptive changes in the climate, particularly, increase in the atmospheric temperature. CO_2 is essentially the waste generated out of the primary energy use. The carbon intensity of the primary energy use also reflects the mix of the energy in the social and economic production system, in particular, the share of the energy from renewable or non-renewable sources, and—among the non-renewable sources—the share of highly carbon-intensive sources such as coal and less carbon-intensive sources such as natural gas (O'neill and Chen 2002).

Following Kim and Strobino (1984), the change in the primary energy use (E) over a period of time can be decomposed into the change in the size of the population, change in the real gross domestic product per capita and the change in the energy intensity of the gross domestic product:

$$\begin{aligned} E_2 - E_1 &= (P_2 - P_1) * A_1 * R_1 + P_1 * (A_2 - A_1) * R_1 + P_1 * A_1 * (R_2 - R_1) \\ &\quad + (P_2 - P_1) * (A_2 - A_1) * R_1 + (P_2 - P_1) * A_1 * (R_2 - R_1) \\ &\quad + P_1 * (A_2 - A_1) * (R_2 - R_1) + (P_2 - P_1) * (A_2 - A_1) * (R_2 - R_1) \\ &= \Lambda + \mathrm{B} + \Gamma + \Lambda\mathrm{B} + \Lambda\Gamma + \mathrm{B}\Gamma + \Lambda\mathrm{B}\Gamma \end{aligned} \tag{11.3}$$

The first three terms on the right-hand side of Eq. (11.3) represent the main effects; the next three terms represent first-order or two-way interactions, while the last term represents second-order or three-way interactions. For example, the first term on the right of Eq. (11.3) shows the contribution of the change in population size to the change in the primary energy use when the real gross domestic product per capita and energy intensity of the gross domestic product remain unchanged. Similarly, the fourth term on the right of the Eq. (11.3) shows the contribution of simultaneous change in the size of the population and the change in the real gross domestic product per capita to the change in the primary energy use when there is no change in the energy intensity of the real gross domestic product. Equation (11.3) thus helps in understanding the relative contribution of the change in the size of the population, change in the affluence and transition in the technology to the change in the primary energy use.

Following the same argument, the change in the CO_2 emission over time may be decomposed in the following manner:

$$\begin{aligned} C_2 - C_1 = & (P_2 - P_1) * A_1 * R_1 * W_1 + P_1 * (A_2 - A_1) * R_1 * W_1 \\ & + P_1 * A_1 * (R_2 - R_1) * W_1 + P_1 * A_1 * R_1 * (W_2 - W_1) \\ & + (P_2 - P_1) * (A_2 - A_1) * R_1 * W_1 + (P_2 - P_1) * A_1 * (R_2 - R_1) * W_1 \\ & + (P_2 - P_1) * A_1 * R_1 * (W_2 - W_1) + P_1 * (A_2 - A_1) * (R_2 - R_1) * W_1 \\ & + P_1 * (A_2 - A_1) * R_1 * (W_1 - W_2) + P_1 * A_1 * (R_2 - R_1) * (W_2 - W_1) \\ & + (P_2 - P_1) * (A_2 - A_1) * (R_2 - R_1) * W_1 \\ & + (P_2 - P_1) * A_1 * (R_2 - R_1) * (W_2 - W_1) \\ & + P_1 * (A_2 - A_1) * (R_2 - R_1) * (W_2 - W_1) \\ & + (P_2 - P_1) * (A_2 - A_1) * (R_2 - R_1) * (W_2 - W_1) \end{aligned} \tag{11.4}$$

or

$$\begin{aligned} C_2 - C_1 = & \alpha + \beta + \gamma + \delta + \alpha\beta + \alpha\gamma + \alpha\delta + \beta\gamma + \beta\delta + \gamma\delta + \alpha\beta\gamma \\ & + \alpha\beta\delta + \alpha\gamma\delta + \beta\gamma\delta + \alpha\beta\gamma\delta \end{aligned} \tag{11.5}$$

Application of Eqs. (11.3) and (11.5) to India has been carried out for two time periods: 1990–2015 and 2015–2030. The period 1990–2015 is the period of the Millennium Development Agenda (United Nations 2000), while the period 2015–2030 is the period of the 2030 Sustainable Development Agenda (United Nations 2015a, b, c). The analysis for the period 1990–2015 shows the historical perspective of the change in the size of the population, change in the per capita consumption or the change in the level of affluence and the change in the technology on the environment in terms of the change in the primary energy use and CO_2 emission. The analysis for the period 2015–2030, on the other hand, projects the future perspective of the projected change in the population size, per capita consumption or the level of affluence and the projected advancement in technology on the projected change in primary energy use and CO_2 emission. In other words, the analysis for the period 1990–2015 tells what has happened in the country in the past, whereas the analysis for the period 2015–2030 tells what is likely to happen in the near future. We build three scenarios for the period 2015–2030.

The first is the Business as Usual scenario. This scenario is based on forecasting the change in the primary energy use and CO_2 emission, change in the size of the population and the change in the per capita consumption on the basis of the trend during 1990–2015 through time series modelling. In the time series modelling, the past provides a model for the future. The Expert Modeller of the Forecasting module of the SPSS software package was used for building the forecasting model (IBM *no date*). The Expert Modeller attempts to automatically identify and estimate the best-fitting ARIMA or exponential smoothing model, thereby eliminating a great deal of the trial and error in modelling time series.

The second scenario is the New Policy Initiative scenario (International Energy Agency 2015). This scenario takes into account a number of policy initiatives that have been taken by India to influence the primary energy use and associated CO_2 emission. Finally, the third scenario may be termed as the Indian Vision case which takes into consideration an accelerated realisation of key policy targets notably Make in India campaign which is directed towards putting industry at the heart of India's growth model (International Energy Agency 2015).

11.3 Data

The analysis is based primarily on the data on primary energy use and CO_2 emission in the country available through EnerData, which is an independent research and consulting organisation on global oil, gas, coal, power, renewable and carbon markets. The organisation provides consistent estimates of total primary energy used, primary energy used per capita, energy intensity of the gross domestic product and total carbon dioxide emission for 186 countries of the world, including India, for different years of the period 1990 through 2015. These estimates have been used in the present analysis. In addition, estimates of population size are derived from the population prospects prepared by the United Nations Population Division in 2017 (United Nations 2017).The population enumerated at different population censuses in the country has not been used in the present analysis because of two counts. First, there is some under counting of the population at all population censuses. For example, the post-enumeration survey carried out after the latest 2011 population census in India suggests that there was a net omission of 23 persons for every 1000 enumerated persons in the country (Government of India 2014). Second, estimates of population from the population census are available at an interval of 10 years only. On the other hand, the official population projections based on the 2011 are not yet available and population projections based on 2001 population census are limited up to the year 2026 only.

11.4 Historical Perspective

It is estimated that, between 1990 and 2015, CO_2 emission in India increased by more than four times from 517 Mt in 1990 to 2166 Mt in 2015 or an increase of 1649 Mt over a period of 25 years (Table 11.1). During this period, the primary energy use increased from 307 Mtoe in 1990 to 882 Mtoe in 2015, or an increase of 575 Mtoe. The increase in the CO_2 emission and the increase in the primary energy use has been associated with an increase in per capita income from around 1580 in 1990 to 5142 in 2015 at 2005 US$ purchasing power parity. Finally, the population of the country is estimated to have increased from 870 million in 1990 to 1309 million in 2015 so that the per capita primary energy used increased from 352 Koe in 1990 to 674 Koe in 2015 while the carbon intensity of the primary energy use increased from 1.686 in 1990 to 2.456 in 2015. Finally, the energy intensity of the gross domestic product decreased from 0.223 in 1990 to 0.131 in 2015.

Table 11.2 decomposes the increase in CO_2 emission during 1990–2015 into the increase attributed to population growth, increase in per capita gross domestic product, increase in the carbon intensity of primary energy use and decrease in the energy intensity of the gross domestic product. Similarly, the increase in the primary energy use during the period 1990–2015 has been decomposed into the increase attributed to population growth, increase in per capita per capita gross domestic product and decrease in the energy intensity of the gross domestic product. Table 11.2 suggests that the primary contributor to the increase in the primary energy use and associated CO_2 emission during the period under reference has been the increase in the gross domestic product per capita or the increase in the affluence. However, the contribution of population growth to the increase in the primary energy use and in the CO_2 emission has also been quite significant. Moreover, the direct contribution of the decrease in the energy intensity of the gross domestic product to the increase in the primary energy use has been too small to compensate for the direct contribution of the increase in the gross domestic product per capita and the increase in population size. Similarly, the direct contribution of the decrease in the energy intensity of the

Table 11.1 CO_2 emission, primary energy use, gross domestic product and population during the period 2015–2030 in India

Particulars	1990	2000	2010	2015
Population (million)	870	1053	1231	1309
Gross domestic product (Billion 2005 US$ ppp)	1375	2360	4878	6731
Gross domestic product per capita (2005 US$ ppp)	1580	2241	3963	5142
Primary energy use (Mtoe)	307	441	693	882
Energy intensity of gross domestic product	0.223	0.187	0.142	0.131
CO_2 emission (Mt)	517	899	1576	2166
Carbon intensity of energy use (Mt)	1.69	2.04	2.28	2.46

Source Author's calculations

Table 11.2 Decomposition of the increase in the primary energy use and CO_2 emission in India during the period 1990–2015

Contribution to the increase in primary energy increase and CO_2 emission	Primary energy use (Mtoe)	CO_2 emission (Mt)
Total increase	575	1649
Direct contribution	720	1449
Population (*P*)	155	261
Gross domestic product per capita (*A*)	691	1166
Energy intensity of gross domestic product ®	−126	−213
Carbon intensity of primary energy use (*W*)		236
Second order contribution	0	553
P and *A*	349	588
P and *R*	−64	−108
P and *W*		119
A and *R*	−285	−481
A and *W*		532
R and *W*		−97
Third order contribution	−144	−243
P and *A* and *R*	−144	−243
P and *A* and *W*		268
P and *R* and *W*		−49
A and *R* and *W*		−220
Fourth order contribution		−111
P and *A* and *R* and *W*		−111

Source Author's calculations

gross domestic product to the increase in CO_2 emission has also been too small to compensate for the direct contribution of the increase in the per capita gross domestic product, population size and the carbon intensity of the primary energy use. This means that technological advancement in the country has not been sufficient enough to compensate for the negative environmental effects associated with the increase in the affluence and the increase in the size of the population.

One way to analyse the impact of population growth on the increase in the primary energy use and CO_2 emission is to estimate the increase in the primary energy use and CO_2 emission under the condition that there is no population growth during the period under reference. Under this assumption, Table 11.2 suggests that the primary energy use would have increased by only 280 Mtoe while the CO_2 emission would have increased by about 923 Mt. This means that the population growth in India during 1990–2015 was responsible for around 296 Mtoe increase in primary energy use and 726 Mt increase in CO_2 emission. The significance of the environmental effects of population growth in India is obvious.

Table 11.3 Forecast of CO_2 emission, primary energy use, gross domestic product and population during the period 2015–2030 in India

Particulars	2015	2030		
		Business as usual	New policy initiative	India vision
Population (million)	1309	1541	1513	1513
Gross domestic product (Billion 2005 US$ ppp)	6731	13408	13408	13408
Gross domestic product per capita (2005 US$ ppp)	5132	8699	8862	8862
Primary energy use (Mtoe)	882	1665	1440	1482
Energy intensity of gross domestic product	0.131	0.124	0.107	0.111
CO_2 emission (Mt)	2166	5138	3744	3865
Carbon intensity of energy use (Mt)	2.46	3.09	2.60	2.61

Source Author's calculations

11.5 Future Prospects

Table 11.3 presents the future prospects of the increase in primary energy use and CO_2 emission and the components of the increase under three different scenarios as discussed above. Under the Business as Usual case, the primary energy use is forecasted to increase to 1665 Mtoe by 2030 whereas the CO_2 emission is forecasted to increase to 5138 Mt. In the New Policy Initiative case, the primary energy use is forecasted to increase to 1440 Mtoe whereas the CO_2 emission is forecasted to increase to 3744 Mt. Finally, in the India Vision case, the primary energy use is forecasted to increase to 1482 Mtoe whereas the CO_2 emission is forecasted to increase to 3865 Mt. On the other hand, the Business as Usual case suggests that the population of the country will increase to 1541 million by the year 2030 whereas the medium variant of population projection prepared by the United Nations Population Division suggests that India's population will increase to 1513 million by 2030. Finally, under the Business as Usual case, the gross domestic product of the country is likely to increase to 13,408 at 2005 US$ purchasing power parity. This means that the energy intensity of the gross domestic product is likely to increase to 0.110, whereas the carbon intensity of the primary energy use is likely to increase to 3.1.

The decomposition of the increase in the primary energy use and increase in CO_2 emission during the period 2015–2030 under three different growth scenarios is presented in Table 11.4. If it is assumed that there will be no increase in the population during the period 2015–2030 then the primary energy use will increase to 532, 364 and 400 Mtoe, respectively, under the Business as Usual, New Policy Initiatives and India Vision scenarios, while the CO_2 emission will increase to 2198, 1073 and 1178

Table 11.4 Decomposition of the increase in the primary energy use and CO_2 emission in India during the period 1990–2015

Contribution to the increase in primary energy use and CO_2 emission	Primary energy use (Mtoe)			CO_2 emission (Mt)		
	Business as usual	New policy initiative	India vision	Business as usual	New policy initiative	India vision
Total increase	783	558	600	2972	1578	1699
Direct contribution	720	616	637	2324	1641	1700
Population (*P*)	156	137	137	384	337	337
Gross domestic product per capita (*A*)	610	638	638	1498	1567	1567
Energy intensity of gross domestic product (*R*)	−46	−159	−138	−113	−390	−338
Carbon intensity of primary energy use (*W*)				555	127	134
Second-order contribution	68	−41	−22	621	−10	43
P and *A*	108	99	99	266	244	244
P and *R*	−8	−25	−21	−20	−61	−53
P and *W*				98	20	21
A and *R*	−32	−115	−100	−78	−282	−245
A and *W*				384	92	97
R and *W*				−29	−23	−21
Third-order contribution	−6	−18	−16	29	−50	−41
P and *A* and *R*	−6	−18	−16	−14	−44	−38

(continued)

Table 11.4 (continued)

Contribution to the increase in primary energy use and CO_2 emission	Primary energy use (Mtoe)			CO_2 emission (Mt)		
	Business as usual	New policy initiative	India vision	Business as usual	New policy initiative	India vision
P and *A* and *W*				68	14	15
P and *R* and *W*				−5	−4	−3
A and *R* and *W*				−20	−16	−15
Fourth-order contribution						
P and *A* and *R* and *W*				−4	−3	−2

Source Author's calculations

Mt, respectively. This means that increase in population during the 15 years between 2015 and 2030 will account, respectively, for 251, 194 and 200 Mtoe increase in the primary energy use and 774, 505 and 521 Mt increase in the CO_2 emission in the country under the three different growth scenarios.

11.6 Discussions and Conclusions

The analysis presented here highlights the impact of the past as well as future population growth in India on the sustainability of the environment in terms of the increase in the primary energy use and associated CO_2 emission. Protection of the environment is one of the three dimensions of sustainable development as defined by the United Nations in the 2030 agenda for sustainable development. The substantial impact of population growth on both increase in the primary energy use and the resulting increase in the CO_2 emission in the country in the past as well as the likely contribution of the future population growth to the projected increase in primary energy use and resulting CO_2 emission suggests that due considerations must be given to issues and concerns related to population transition while customising the United Nations 2030 Sustainable Development Agenda to the Indian context. The population of the country is still in the middle stage of transition so that population effects of environmental sustainability in the country are quite substantial as is evidence from the present analysis. This means that accelerating the pace of population transition should be one of the sustainable development agenda that is specific to the country.

The analysis also suggests that although the direct contribution of the population growth to the increase in the primary energy use and resulting CO_2 emission may not be large, yet the second-order effects of population growth in conjunction with the change in the per capita gross domestic product or the change in the affluence are quite substantive. This means that a substantial proportion of the contribution of the increase in affluence to the increase in the primary energy use and the resulting increase in the emission of greenhouse gases including CO_2 can be compensated by accelerating the pace of population transition in the country. This observation is important as the increase in affluence measured in terms of gross domestic product per capita is widely recognised as an indicator of development and improvement in the quality of life.

Finally, the analysis shows that technology advancement had and is likely to have only a limited impact on the environmental concerns in India. The decrease in the primary energy use and the resulting decrease in CO_2 emission attributed to the improvement in the technology used in the social and economic production system have been found to be too small to compensate for the increase in the primary energy use and CO_2 emission as a result of population growth and the increase in affluence. The prospects of rapid advancement in technology in India appear to be remote because of the heavy cost associated with technological innovations and technological research. As such, the case of accelerating population transition in the context of sustainable development in India becomes even stronger.

References

Commoner B, Corr M, Stamler PJ (1971) The closing circle: nature, man, and technology. Knopf, New York

Dietz T, Rosa E (1997) Effects of population and affluence on CO2 emissions. Proc Natl Acad Sci 94:175–179. USA

Ehrlich PR (1968) The Population Bomb. Ballantine Books, New York

Government of India (2014) Census of India 2011. Report on Post Enumeration Survey. Registrar General and Census Commissioner, India, New Delhi

IBM (no date) IBM SPSS forecasting. Armonck, NY, IBM Corporation

International Energy Agency (2015) India energy outlook. Paris, International Energy Agency

Kaya Y (1990) Impact of carbon dioxide emissions control on GNP growth: interpretation of proposed scenarios. Paper presented to the IPCC Energy and Industry Subgroup. Paris, Response Strategies Working Group

Kim YJ, Strobino DM (1984) Decomposition of the difference between two rates with hierarchical factors. Demography 21(3):361–372

O'neill BC, Chen BS (2002) Demographic determinants of household energy use in the United States. In: Lutz W, Prskawetz A, Saunderson WC (Eds) Population and Environment: Methods of Analysis. New York, Population Council. Supplement to Popul Dev Rev 28: 53–88

Stern PC, Young OR, Druckman D (1992) Global environmental change: understanding the human dimensions. National Academy Press, Washington DC

United Nations (1994) Report of the international conference on population and development. New York, United Nations. No. A/Conf.171/13

United Nations (2000) United Nations millennium declaration. New York, United Nations. No. A/Res/55/2

United Nations (2015a) Transforming the world: the 2030 agenda for sustainable development. New York, United Nations. No. A/RES/70/1

United Nations (2015b) Press kit for the sustainable development summit 2015: time for global action for people and planet. New York, United Nations

United Nations (2015c) World population prospects: the 2015 revision. New York, United Nations, Department of Economic and Scientific Affairs. Population Division

United Nations (2017) World population prospects. The 2017 revision. Department of Economic and Social Affairs. Population Division, New York

The manufacturer's authorised representative in the EU is Springer Nature Customer Service Centre GmbH, Europaplatz 3, 69115 Heidelberg, Germany. If you have any concerns regarding our products, please contact ProductSafety@springernature.com

Printed and bound by CPI Group (UK) Ltd, Croydon, CR0 4YY
25/09/2026
02222900-0008